INTEGRATING DEEP LEARNING ALGORITHMS TO OVERCOME CHALLENGES IN BIG DATA ANALYTICS

Green Engineering and Technology: Concepts and Applications

Series Editors: Brujo Kishore Mishra, GIET University, India and Raghvendra Kumar, LNCT College, India

The environment is an important issue these days for the whole world. Different strategies and technologies are used to save the environment. Technology is the application of knowledge to practical requirements. Green technologies encompass various aspects of technology that help us reduce the human impact on the environment and creates ways of sustainable development. This book series will enlighten the green technology in different ways, aspects, and methods. This technology helps people to understand the use of different resources to fulfill needs and demands. Some points will be discussed as the combination of involuntary approaches, government incentives, and a comprehensive regulatory framework that will encourage the diffusion of green technology; underdeveloped countries and developing states of small islands require unique support and measure in order to promote green technologies.

Machine Learning and Analytics in Healthcare Systems
Principles and Applications
Edited by Himani Bansal, Balamurugan Balusamy, T. Poongodi, and Firoz Khan KP

Convergence of Blockchain Technology and E-Business:
Concepts, Applications, and Case Studies
Edited by D. Sumathi, T. Poongodi, Bansal Himani, Balamurugan Balusamy, and Firoz Khan K P

Big Data Analysis for Green Computing
Concepts and Applications
Edited by Rohit Sharma, Dilip Kumar Sharma, Dhowmya Bhatt, and Binh Thai Pham

Handbook of Sustainable Development Through Green Engineering and Technology
Edited by Vikram Bali, Rajni Mohana, Ahmed Elngar, Sunil Kumar Chawla, and Gurpreet Singh

Integrating Deep Learning Algorithms to Overcome Challenges in Big Data Analytics
Edited by R. Sujatha, S. L. Aarthy, and R. Vettriselvan

For more information about this series, please visit: https://www.routledge.com/Green-Engineering-and-Technology-Concepts-and-Applications/book-series/CRCGETCA

INTEGRATING DEEP LEARNING ALGORITHMS TO OVERCOME CHALLENGES IN BIG DATA ANALYTICS

Edited by
R. Sujatha, S. L. Aarthy, and R. Vettriselvan

CRC Press
Taylor & Francis Group
Boca Raton London New York

CRC Press is an imprint of the
Taylor & Francis Group, an **informa** business

First edition published 2022
by CRC Press
6000 Broken Sound Parkway NW, Suite 300, Boca Raton, FL 33487-2742

and by CRC Press
2 Park Square, Milton Park, Abingdon, Oxon, OX14 4RN

© 2022 Taylor & Francis Group, LLC

CRC Press is an imprint of Taylor & Francis Group, LLC

Library of Congress Cataloging-in-Publication Data
Names: Sujatha, R. (Computer science professor), editor. I Aarthy, S. L., editor. I Vettriselvan, R., editor.
Title: Integrating deep learning algorithms to overcome challenges in big data analytics / edited by R. Sujatha, S.L. Aarthy, R. Vettriselvan.
Description: Boca Raton : CRC Press, 2022. I Series: Green engineering and technology: concepts and applications I Includes bibliographical references and index.
Identifiers: LCCN 2021016673 (print) I LCCN 2021016674 (ebook) I ISBN 9780367466633 (hbk) I ISBN 9781032104461 (pbk) I ISBN 9781003038450 (ebk)
Subjects: LCSH: Machine learning--Industrial applications. I Big data. I Artificial intelligence--Industrial applications. I Algorithms.
Classification: LCC Q325.5 .I475 2022 (print) I LCC Q325.5 (ebook) I DDC 006.3/1--dc23
LC record available at https://lccn.loc.gov/2021016673
LC ebook record available at https://lccn.loc.gov/2021016674

ISBN: 978-0-367-46663-3 (hbk)
ISBN: 978-1-032-10446-1 (pbk)
ISBN: 978-1-003-03845-0 (ebk)

DOI: 10.1201/9781003038450

Typeset in Times
by MPS Limited, Dehradun

Contents

Preface

Data science revolves around two giants: Big Data analytics and Deep Learning. Both public and private sectors accumulate huge data of specific domains that hold useful information. Big Data analytics is the process of mining and extracting patterns from hefty data sets for prediction and, in turn, to make a beneficial decision. Yet, it is too challenging a procedure because of the non-uniformity of the data; because streaming data comes at a gushing speed; because the input source is highly distributed; and obviously because clarity is missing in input data. In 1957, Rosenblatt worked on the principle that makes a machine learn and classify the human way. This system requires two basic things, namely, the ability to recognize the complex system, and the ability to perform which requires voluminous amounts of data and, in turn, to provide information. This is the basis for Deep Learning, which works as the basis of neural networks with a number of layers of nodes between entry and exit. It is a well-known fact that data is expanding at a greater rate and by 2020 it will reach 40 zettabytes of data. Evidently, it is going to be challenging to handle and retrieve useful information from this huge data set. The optimization of Deep Learning algorithms over Big Data to accommodate the challenges is a great area of research. The chapters of this book act as a great support to address this issue in several areas. A thorough analysis of Big Data along with Deep Learning is mandatory to ensure flawless analysis or classification over uncategorized data. The research pertaining to addressing these challenges of Deep Learning over Big Data is in a preliminary and evolutionary phase. It is inevitable that it will provide insight into the advent of Deep Learning in all the fields and it is going to be a potential research area in data science that is growing at an extraordinary rate. Irrespective of the domain, Big Data analytics, along with the Deep Learning part of Machine Learning, makes the study interesting. IBM has predicted that the demand for data scientists will soar by 2020. It is clear that all target audiences will benefit by building a framework-based solution, which caters to each based on the requirements and provides the best platform irrespective of the domain. It helps in taking effective decisions with available huge generated data along with effective algorithms to solve the problem. For instance, in healthcare, various stakeholders are patients, medical practitioners, hospital operators, pharma and clinical researchers, and healthcare insurers. Various tasks from data integration, searching, processing, Machine Learning, streamed data processing, and visual data analytics are various tools that make a highly user-friendly healthcare environment.

This book will bring forth, in a precise and concise manner, the details about all the right and relevant technologies and tools to simplify and streamline the formation of Big Data, along with Deep Learning System architects and designers, the data coders, statisticians, business people, researchers, and others.

R. Sujatha
S. L. Aarthy
R. Vettriselvan

Editors

R. Sujatha is an Associate Professor at the School of Information Technology and Engineering, Vellore Institute of Technology. She earned a Ph.D. in data mining at Vellore Institute of Technology in 2017, an ME in computer science at Anna University in 2009, with the ninth rank in university, a Master's degree in Financial Management at Pondicherry University in 2005, and a BE in computer science at Madras University in 2001. Dr. Sujatha has 17 years of teaching experience. She has organized and attended a number of workshops and faculty development programs. Dr. Sujatha is actively involved in the growth of Vellore Institute through various committees at both the academic and administrative levels and guides undergraduate, postgraduate, and doctoral students. She gives technical talks and acts as an advisory, editorial member, and technical committee member in various symposiums. Dr. Sujatha authored *Software Project Management for College Students* as well as research articles in high-impact journals. The Institution of Green Engineers awarded her the IGEN Women Achiever in 2021 in the category of future computing. Dr. Sujatha is interested in exploring different places to learn about various cultures and people. Her areas of interest include data mining, Machine Learning, software engineering, soft computing, Big Data, Deep Learning, and blockchain.

S. L. Aarthy is an Associate Professor at the School of Information Technology and Engineering, Vellore Institute of Technology. She earned a Ph.D. degree in medical image processing at Vellore Institute of Technology in 2018, an ME in computer science at Anna University in 2010, and a BE in computer science at Anna University in 2007. Dr. Aarthy has 11 years of teaching experience. Her research interest includes image processing, soft computing, and data mining. She has published approximately 20 papers in reputed journals, and she guides undergraduate, postgraduate, and doctoral students. She is a life member of CSI and IEEE, as well as various institute committees.

R. Vettriselvan is an Assistant Professor at AMET Business School, AMET (Deemed to be University), Chennai. He earned a BA in economics at Madurai Kamaraj University, an MBA at Anna University, an MPhil in research and development, and a Ph.D. at Gandhigram Rural Institute–Deemed University. Dr. Vettriselvan received an ICSSR Doctoral Fellowship, ICSSR, New Delhi; a GRI Fellowship; and a Post-Graduate Diploma in personnel management and industrial relations at Alagappa University, Karaikudi. He specializes in human resource management and marketing. Dr. Vettriselvan has published 5 books and 65 research articles in SCOPUS, UGC, referred international, national peer-reviewed journals, and conference volumes. He has presented more than 60 research articles at national and international conferences in India, Zambia, Malawi, and the United States. He received a travel grant award in 2015 from the Population Association of America, USA, to present a research article in California. Dr. Vettriselvan has received recognition for "best paper", "best paper presenter", "best young faculty", "bright educator", "best academician of the year (male)", and "most promising educator in higher education" across India. He is an editorial and review board member for a number of peer-reviewed journals. He is guiding three Ph.D. research scholars and has guided 30 MBA and 86 undergraduate projects. He has experience in NBA and NAAC acceptance documentation processes. He has 6 years of experience as a manager in human resources, NEST Abroad Studies Academy Private Limited, Madurai; lecturer and head of the department, School of Commerce and Management, DMI–St. Eugene University, Zambia; assistant professor, AMET Business School, AMET University, India; and lecturer at the School of Commerce and Management and Coordinator, Department of Research and Publication, DMI–St. John the Baptist University. He has been a review member for the National Council for Higher Education, Malawi (February 2020 to June 2020) to review the accreditation process.

Contributors

S. L. Aarthy
School of Information Technology and
Engineering
Vellore Institute of Technology
Vellore, India

D. Aju
School of Computing Science and
Engineering
Vellore Institute of Technology
Vellore, India

N. Anand
School of Computing Science and
Engineering
Vellore Institute of Technology
Vellore, India

P. Anbumani
Department of Computer Science and
Engineering
V.S.B. Engineering College
Karur, India

Vanmathi C
School of Information Technology and
Engineering
Vellore Institute of Technology
Vellore, India

D. Deepa
School of Computing Science and
Engineering
Vellore Institute of Technology
Vellore, India

Erapaneni Gayatri
School of Information Technology and
Engineering
Vellore Institute of Technology
Vellore, India

D. Helen
AMET University
Chennai, India

S. Karthi
Department of Computer Science and
Engineering
V.S.B. Engineering College
Karur, India

P. Kasthurirengan
Department of Computer Science and
Engineering
V.S.B. Engineering College
Karur, India

Anish M. Lal
School of Computing Science and
Engineering
Vellore Institute of Technology
Vellore, India

P. Latha
Department of Computer Science and
Engineering
V.S.B. Engineering College
Karur, India

R. Mangayarkarasi
School of Information Technology and
Engineering
Vellore Institute of Technology
Vellore, India

S. Nathiya
School of Information Technology and
Engineering
Vellore Institute of Technology
Vellore, India

C. Padmapriya
Rose Mary College of Arts and Science
Tirunelveli, India

M. Parthiban
Department of Information Technology
V.S.B. Engineering College
Karur, India

S. Prabu
School of Computing Science and
    Engineering
Vellore Institute of Technology
Vellore, India

M. Sangeetha
Department of Computer Science and
    Engineering
V.S.B. Engineering College
Karur, India

A. Suganya
School of Information Technology and
    Engineering
Vellore Institute of Technology
Vellore, India

R. Sujatha
School of Information Technology and
    Engineering
Vellore Institute of Technology
Vellore, India

1 A Study on Big Data and Artificial Intelligence Techniques in Agricultural Sector

D. Helen and C. Padmapriya

CONTENTS

DOI: 10.1201/9781003038450-1

1.1   INTRODUCTION

Agriculture plays a vital role in the overall development of a country's economy. Agriculture is the major source of livelihood (Guruprasad et al., 2019). To ensure the financial development of a country, it is necessary to monitor and estimate crop production (Shah & Shah, 2019). The main aim of the country is to increase crop yield using minimal resources (Kumar et al., 2015). The yield prediction is most important for universal food production. The accurate crop prediction and the timely report reinforce the overall food security. The crop yield prediction helps the government in planning for manufacturing, supply, and utilization of the food. The major issue for agricultural development is an accurate yield prediction for the number of crops involved in the production.

Big Data and Artificial Intelligence (AI) is an emerging technology in the agricultural sector, which automates the agricultural process. In the agricultural field, AI techniques are applied in three major areas: (1) Artificial Robots, which harvest crops faster and in high volume; (2) Deep Learning and Computer Vision techniques, which monitor the health of crops and soil; and (3) Predictive Analysis Method, which predicts environmental changes such as temperature, rainfall, etc.

AI techniques work efficiently in complex relationships between Input and Output variables (Jain et al., 2017). AI techniques depend on the semi-parametric and non-parametric structures, and the justification is based on accurate prediction (Breiman, 2001). Machine Learning, Artificial Neural Networks (Fortin et al., 2011; Liu et al., 2001), Regression Trees, and Support Vector Machines (Jaikla et al., 2008) are the popular AI techniques used for crop yield prediction.

1.1.1   The Life Cycle of Agriculture

Soil Preparation: This is the first stage of farming where farmers sow the seeds in the soil. This process involves breaking up huge soil clumps and removing wreckage such as sticks, rocks, and roots. Fertilizers and organic matter are added according to the type of crop.

Sowing: At this stage, climate conditions play an important role. The distance between two seeds and the depth for planting seeds is necessary while sowing the seeds.

Adding Manures and Fertilizers: Soil fertility is an important factor that helps farmers to grow nutritious and healthy crops. Fertilizers are chemical substances with the composition of nitrogen, phosphorus, and potassium that are added to the soil to increase crop productivity. Crop yield can be increased by adding manure and fertilizers.

Irrigation: Humidity and soil moisture can be maintained at this stage. Watering the crops plays an important role here. Underwatering and overwatering can damage the growth of crops.

Weeding: Weeds are the unwanted plants that grow along with the main crops. Weeding plays a necessary role in agriculture because the presence of weeds decreases crop yield, reduces crop quality, and increases production cost.

Harvesting: In this phase, ripe crops are collected from the fields. A lot of laborers are needed during this activity. Harvesting also includes post-harvest handling such as cleaning, sorting, packing, and cooling.

Threshing: This is the process of removing grains from the straw and chaff. This operation can be carried out manually or through machines. Threshing can be done by three methods that include rubbing, impact, and stripping.

Storage: Foodgrains obtained after harvesting should be dried to remove moisture. The products are stored in such a way as to guarantee food security. Grains are stored in silos. This phase also includes the packing and transportation of crops (Figure 1.1).

1.2 THE ROLE OF BIG DATA IN THE AGRICULTURAL SECTOR

Big Data plays a prominent role in the agricultural sector. Big Data is a combination of technology and analytics that can collect a huge amount of both structured and unstructured data. It compiles and processes these data effectively to assist in decision-making (Sonka, 2014). To process these large amounts of data, advanced tools are required. Real-time data analytics and automated processing are done with Big Data. In order to implement Big Data successfully, many techniques are used,

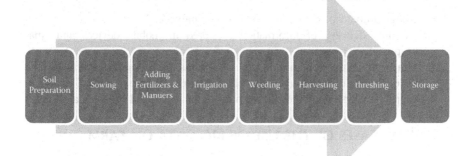

FIGURE 1.1 The life cycle of agriculture.

such as predictive analytics, machine learning, time series analysis, classification and clustering, data mining, regression analytics, etc.

More advancement in the technology of Big Data can establish a smart agricultural system. Agriculture is rapidly moving from traditional methods to these modern tools and technology. A farming process would be simpler and better with the help of Big Data. Big Data can solve complex problems in agricultural systems. Massive data are collected through various kinds of control devices, drones, satellites, and sensors. These data are analyzed and used to plan for better crop production. Farmers can face critical problems regarding decision-making (Sonka, 2014). Using Big Data analytics, farmers can make predictions and appropriate decisions with the data drawn from the preceding years' rainfall and climate conditions to avoid crop failure. Big Data not only creates smart farming but also influences the supply and marketing chain. Thus, Big Data helps the farmers make appropriate decisions, such as when to irrigate the field, as well as weather and crop health predictions. As there are labor shortages in agriculture, Big Data analysis can reduce the need for physical manpower. The end result of Big Data is to give better results at the right time from the gathered data. This advancement in technology increases crop production and the economic condition of the country.

1.2.1 OVERALL CHARACTERISTICS OF BIG DATA APPLICABLE TO THE AGRICULTURAL SECTOR

- **Volume:** Huge amount of data is stored. Massive data are collected through various kinds of control devices, drones, satellites, and sensors.
- **Velocity:** Real-time analysis and decision-making are carried out using Big Data to predict weather forecasting, fertilizer requirements, pest infestation, water availability, nutritional status of the soil, and also can send alerts in emerging situations at an optimal time from the data which are collected.
- **Variety:** A variety of file formats with different sizes from various devices is collected. The collected data may be text, images, audio, or videos. Appropriate decisions are carried out by connecting all these data points.

1.2.2 THE PROCESSING STEPS OF BIG DATA IN AGRICULTURE

1. A huge amount of data is combined and analyzed.
2. Real-time delivery of information is stored in mobile devices and smart equipment.
3. Data from weather conditions, soil conditions, GPS mapping, fertilizer/pesticide use, water resources, and field characteristics are analyzed.
4. Appropriate decisions are carried out by connecting all these data points.

1.3 SOME CASES OF THE USE OF BIG DATA ON FARM

1.3.1 TO EVADE FOOD SCARCITY OF THE GROWING POPULATION

A growing population is one of the most common problems of the world. The government takes many steps to solve the problems faced by overpopulation.

One of the foremost challenges is feeding the growing population. This can be achieved by increasing the yield of present farmlands. Farmers can face critical problems regarding decision-making. Using Big Data analytics, they can conclude predictions and make appropriate decisions with the help of data drawn from the preceding years' rainfall and climate conditions to avoid crop failure. Thus, the farmers can increase the yield of crop production using these advanced technologies.

1.3.2 Managing Pesticides and Farm Equipment

Pesticides play an important role in agriculture. Pesticides include chemicals or natural products, which are used to get rid of different pest organisms such as insects, weeds, plant diseases, rodents, and mollusks. Farmers can manage and use the right amount of pesticides according to the soil type using Big Data. This can avoid massive chemicals in food production. Big Data increases the profit for the farmers as the crops are not getting destroyed by insects and weeds. Farming equipment is integrated with sensors so that the farmers can manage tractor availability, fuel refill alerts, and service due dates.

1.3.3 Supply Chain Management

Every year, food produced for human feeding is lost or wasted. The world struggles to bridge the gap between supply and demand. To solve these problems, food delivery cycles from producers to markets must be minimized. Big Data can achieve supply chain improvement effectively by tracking and regulating the delivery truck routes. Farmers and all stakeholders are flexible in choosing their business partners and modern technology for the sake of production and development. Multistakeholders can easily collaborate and function together effectively in an agroecosystem using Big Data. It promotes transparent operations in the agricultural sector. The import and export of agricultural products are monitored using digital technology. Big Data helps farmers and distributors to enhance fleet management.

1.3.4 Yield Prediction and Risk Management

Mathematical models are used to analyze data regarding weather conditions, yield, and chemicals. Thus, Big Data helps the farmers to make appropriate decisions, such as when to irrigate the field, as well as weather and crop health predictions. This modern technology also helps the farmers to plant their seeds with appropriate distance and depth. The historic yield records help the farmers in making appropriate decisions to overcome risk management.

1.4 CHALLENGES FACED BY FARMERS VERSUS AI SOLUTIONS

The traditional methods followed by farmers are not able to satisfy the increasing demand of agricultural products. Some of the challenges faced by farmers are described in the following subsections (Dwivedy, 2011).

1.4.1 FORECASTING WEATHER CONDITIONS

Weather conditions play a major role in the agricultural life cycle. The tremendous increase in deforestation and pollution affects climate change. Hence, improper climate prediction leads to crop loss. Farmers face challenging situations for making proper decisions about sowing seeds and harvesting. The Regression Analysis will help farmers to forecast the weather conditions. This helps the farmers in making proper decisions about sowing the seeds and harvesting.

1.4.2 DECISION-MAKING

Decision-making plays a vital role in the agricultural sector. The predictive analysis helps the farmers to make the correct decision about choosing the right seed for the right area in the following ways:

- The accurate decision for sowing
- Predicting the healthy crop yield
- Recommendation of fertilizer depending on the plant situation

1.4.3 DIAGNOSING DEFECTS IN SOIL AND WEED DETECTION

Insects, weeds, and diseases are the biological factors that affect the yield of crops. Controlling them is a major challenge for farmers. Anomaly Analysis helps to identify the defects in soil, and Image Classification techniques help to identify the weeds on the farm.

1.4.4 NUTRITION TREATMENT

Nutrients such as nitrogen, phosphorous, and potassium found in soil are very essential for the growth of crops. The absence of nutrients can lead to poor-quality crops. Nutrients are very important for increasing crop yield. The Anomaly Detection Mechanism uses an unsupervised learning method to find an appropriate mineral level for growing healthier plants faster.

1.5 AI TECHNIQUES IN AGRICULTURAL SECTOR

1.5.1 MACHINE LEARNING

Machine Learning (ML) is a subfield of Artificial Intelligence (AI). The ML is a tool that turns data into knowledge. The ML algorithms help the system learn from the dataset and accurately predict the result without being programmed explicitly (Okori & Obua, 2011). The ML algorithms receive the input dataset and apply statistical methods to forecast a result (Shastry et al., 2016). The ML methods can automatically discover hidden patterns within complex data. The hidden patterns can be used to forecast the future of the decision-making process. The process of the ML algorithm includes data collection, data pre-processing, training, evaluation, and tune (Snehal & Sandeep, 2014). The ML technique can be used in various applications such as recognizing anomalies, pattern recognition, prediction, neural networks, and so on. Machine Learning is a model of an automated data-processing algorithm.

The Machine Learning algorithm models are:

- **Supervised Machine Learning:** The outputs are labeled, and the inputs are mapped to subsequent outputs.
- **Unsupervised Machine Learning:** The inputs are unlabeled, and the algorithms must discover patterns.
- **Reinforcement Machine Learning:** Similar to a supervised ML, but as an alternative for a labeled output, there are rewards that are given to the algorithm.

1.5.1.1 Supervised Learning

Supervised Learning is a machine learning task that builds an algorithm to learn and map the input for the desired output. The Supervised Learning system learns the labeled training data set and predicts the output of unseen data. In the Supervised Learning process, collected datasets are labeled as training data. The training data contains the pair of input data (vector) and preferred output data (supervisory signal).

In Supervised Learning, the algorithm uses the input variable (X) and an output variable (Y) to learn and map the function of the input to the output.

$$Y = f(X)$$

The mapping function maps the input data (X) that predicts the output variable (Y) for that dataset (Figure 1.2).

The key benefits of a Supervised Learning algorithm are as follows:

- With the Supervised Learning algorithm, knowledge can be used to predict the output of new unnoticed data.
- The knowledge gained by the algorithm can be used for optimizing performance.
- Supervised Learning algorithms are more suitable for real-world problems.

1.5.1.2 Unsupervised Learning

The Unsupervised Learning algorithms are trained using an unlabeled dataset and the models act on information without any guidance. Unsupervised Learning is a self-learning process where the model can discover the unknown patterns from the dataset.

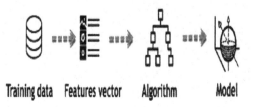

Training data   Features vector   Algorithm   Model

FIGURE 1.2 Supervised algorithm learning phase.

The key benefits of Unsupervised Learning algorithm are as follows:

- The algorithm works on an unlabeled dataset and discovers new patterns.
- The discovered patterns are used to classify elements and uncover the execution between the data.
- The algorithm can identify anomalies and defects in the dataset.

1.5.2 NEURAL NETWORKS

The Neural Network operates like a human brain and its structure is similar to the neuron cells of a human brain. The "neuron" in the Neural Network represents the mathematical functions that organize and classify information based on the specific function.

The typical Neural Network contains three layers: one input layer, one or more hidden layers, and an output layer. The layers are connected by interconnected nodes (Figure 1.3).

The typical Neural Network contains three layers:

- **Input Layer:** The input layer contains the neurons that receive the inputs from the outside world. It does not perform any computation process; it simply forwards the data to the hidden layer.
- **Hidden Layer:** The hidden layer works between the input and output layers. The computation processes are performed in this layer. Generally, the network has zero or multiple hidden layers.
- **Output Layer:** The input processes through the sequence transformation through the hidden layer and the result will be delivered in the output layer. The output layer is responsible for transferring the information to the outside world.

1.5.2.1 Working Process of Neural Network

The Neural Network layers are connected with interconnected nodes and all the nodes are called "perceptron". The perceptron is the essential unit of the Neural Network and contains four major components:

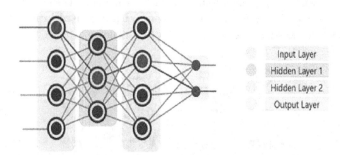

FIGURE 1.3 Neural network architecture.

1. Input
2. Weight and Bias
3. Summation Function
4. Activation Function

Input: The inputs are received from the input layer.

Weights and Bias: After the input variables are fed into the network, a value is randomly chosen and it is assigned as the weight of the input. The weight of each input is important for predicting the output. The bias attributes regulate the activation function curve and help it achieve the desired output.

Summation Function: After the input is assigned to the weight, the product of the corresponding input and weight are taken. Adding all the products together produces the weighted sum. All these activities are done by the summation function.

Activation Function: The activation function maps the weighted sum to the output (Figure 1.4).

The logic behind a perceptron is as follows:

The inputs (X_i) are received from the input layer and multiplied by their weights (W_i). Then the summation function adds the multiplied values and forms a weighted sum. The input of the weighted sum is applied to the activation function (F). The activation function maps the input to and produces the output (Y).

1.5.3 THE EXPERT SYSTEM

The expert system is a consistent and interactive computer-based decision-making system. It solves complex problems using facts and heuristics. An expert system is able to handle challenging decision problems and give solutions according to the domain of knowledge.

1.5.3.1 Components of the Expert System

User Interface: User interface plays a vital role in an expert system. The user interface acts as an interface between the user and the expert systems for communication. It accesses the user input in a readable form and passes it to the interface engine.

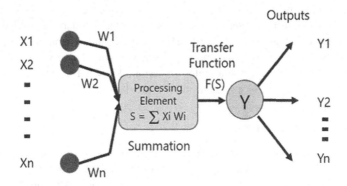

FIGURE 1.4 Working process of neural network.

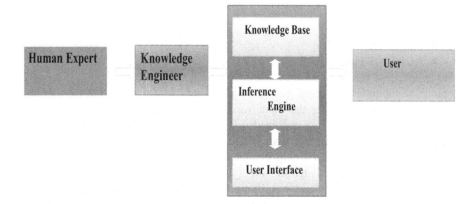

FIGURE 1.5 Expert system.

Interface Engine: The interface engine acts as the brain of the expert system. It solves all critical problems according to the knowledge that it accesses from the knowledge base. It applies facts and rules to solve the users' queries. This component is extremely essential in formulating conclusions.

Knowledge Base: The knowledge base is a storehouse like a large container that holds the facts and knowledge obtained from various experts in a specific field about the problem domain. The success of an expert system mainly depends on highly accurate and exact knowledge (Figure 1.5).

1.5.3.2 The Working Process of the Expert System

The Expert System works as follows:

- Analyze the characteristics of the problem.
- Knowledge engineers and domain experts should work in unity to define the problem.
- Knowledge acquisition is an essential part of the expert system. The main aim is to extract the required knowledge from the human expert and to convert the acquired knowledge into rules. The developed rules are stored in the knowledge base.
- The knowledge is converted into computer-understandable knowledge from the knowledge engineer. The interface engine and reasoning structure are designed by the knowledge engineer to access knowledge whenever needed.
- The knowledge expert integrates uncertain knowledge into the reasoning process to give useful explanations (Figure 1.6).

1.5.4 THE DECISION TREE

The decision tree is the visual representation of decision-making, which falls under the category of supervised learning. It is represented as a treelike model that starts at the root node with two nodes, including the decision node and a leaf node. Decision nodes are used to make any decisions with multiple branches. Leaf nodes

FIGURE 1.6 Working process of expert system.

are the final output of those decisions with no further branches. Based on the given conditions and the answer (yes or no), it further splits the tree into subtrees. It is commonly used in data mining to derive a strategy and to solve regression and classification problems (Figure 1.7).

1.5.4.1 Working Steps of the Decision Tree

Step 1: Start the tree with the root node which encloses the whole data set.
Step 2: Attribute selection measure is used to find the best attributes in the data set.
Step 3: Divide the root node into subtrees, which may contain possible values for the best attributes.
Step 4: The decision tree node is generated with the best attribute.
Step 5: Step 3 is recursively continued to make new decisions. This process is continued until the nodes are further classified. That final node is called the leaf node.

1.5.5 SUPPORT VECTOR MACHINE

A Support Vector Machine (SVM) is a supervised machine learning model, which uses classification algorithms for regression and classification problems. It classifies linear and nonlinear data. Many pattern recognition problems are solved using texture classification. SVM is constructed to work in a multi-dimensional space with two classes by defining a hyperplane to separate the

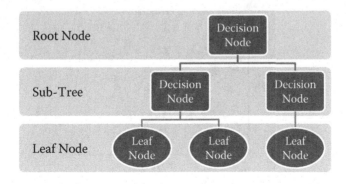

FIGURE 1.7 Decision tree.

two classes. This work is done by increasing the margin from the hyperplane of the two classes. The samples that are closest to the margin are known as support vectors.

The goal of SVM is to find the Maximum Marginal Hyperplane (MMH) by dividing the data sets into classes using the following steps:

1. A hyperplane is generated by SVM iteratively that classifies the classes in the best way.
2. Then the hyperplane is selected, which separates the classes accurately (Figure 1.8).

1.5.6 RANDOM FOREST

Random Forest (RF) is supervised machine learning with classification and regression algorithms. It resembles a decision tree with little improvement. This algorithm creates a forest randomly with several trees. Increasing the number of trees in the forest gives highly accurate results. Multiple decision trees are built in the forest to give the best results and the most stable prediction. The decision tree built in the forest is trained using the bagging method. Randomly, it creates a decision tree for choosing parameters from the data set and building multiple decision trees. Each tree chooses the class. The class that receives the maximum votes is termed the predicted class.

1.5.6.1 Working Steps of an RF

Step 1: The algorithm selects the random samples that are selected from the data set.

Step 2: A decision tree is created for each sample that was selected. The prediction result is estimated for each decision tree.

Step 3: Voting is examined for each predicted result.

Step 4: Final Prediction of the class is estimated from the voted result.

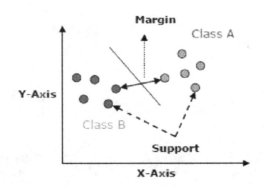

FIGURE 1.8 Support vector machine.

1.6 APPLICATION OF AI IN AGRICULTURE

Agriculture plays a significant role in the economic development of every country. As the world population is increasing tremendously, it is essential to increase agricultural production by introducing modern technology. Some of the AI technology that is implemented in the agricultural sector is summarized below.

1.6.1 IMAGE RECOGNITION

The yield of crop production and crop growth in agriculture is monitored using drones. At each stage of the production cycle, the fields are scanned by the drones. This helps farmers make appropriate decisions from the captured images. They can analyze the health of the crops, soil variation, fungal infestation, and can protect their farms using pest control, as well as take various actions to increase crop production. The AI technology implemented with drones helps farmers to save a lot of time.

1.6.2 DISEASE DETECTION

The images that are captured are used to classify the leaf images into the non-diseased parts and the diseased parts. The diseased parts, which are identified from pre-processing of the image, are sent to a lab for further diagnosis (Figure 1.9) (Kumar et al., 2015).

1.6.3 FIELD MANAGEMENT

Images of various crops are captured and farmers can categorize the fruits by determining how ripe the green fruits are. This helps farmers pick and pack fruits at the correct time. Farmers are able to cultivate their crops by identifying the areas where crops require water, pesticides, and/or fertilizer.

1.6.4 DRIVERLESS TRACTOR

Self-driving vehicles are automated vehicles. It is a driverless vehicle that operates without the presence of humans. These vehicles are operated with hardware and software programming. Agricultural tasks and tillage are done with a high tractive effort at low speed. It provides safer working environments for

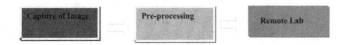

FIGURE 1.9　Disease detection.

farmers. It is programmed in such a way that it can detect their positions and speed, and can avoid complications such as human beings, objects, and animals while performing tasks. Their work is monitored by a supervisor at a control station and can also be operated by remote control from a distance.

1.6.5   WEATHER FORECASTING

The advanced technology of AI helps farmers analyze weather conditions. This helps the farmers plan and sow crop varieties according to climatic conditions. These predicted weather conditions help farmers increase their yield and profits without any risk.

1.6.6   AI AGRICULTURAL BOTS

AI–agriculture bots help the farmers protect their crops from weeds more effectively. AI bots are able to harvest crops very fast and at a higher volume than human laborers. Using computer vision, they are able to monitor weeds and spray them. Thus, AI helps farmers protect their crops from weeds quite effectively.

1.6.7   REDUCTION OF PESTICIDE USAGE

AI technology helps farmers check weeds from the gathered data, and can spray chemicals only at the particular places where weeds are located. This reduces the use of spraying chemicals over whole fields. Thus, the volume of chemicals or pesticides used is comparatively reduced.

1.7   ADVANTAGES OF USING AI IN AGRICULTURE

- AI provides effective ways to yield, harvest, and sell essential crops.
- Implementation of AI helps detect the defective parts of crops and can aid in growing healthy crops.
- Agriculture-based business is improved more effectively using AI in the growth process.
- The crop management process is improved so that many business people can invest in agriculture.
- AI solutions are able to solve all the problems faced by farmers, such as climatic changes, weeds, and the infestation of pests.
- AI technology is able to identify diseased crops quite accurately.
- Fruits and vegetables are monitored frequently so that the production is increased by taking necessary actions at the correct time.
- The nutrient deficiency of the soil is monitored using AI. This aids in healthy crop production.

1.8 CONCLUSION

The agricultural field faces a lot of challenges such as lack of irrigation systems, inability to predict weather conditions, the health of the soil, and the detection of diseases. AI is an emerging technology that helps farmers analyze various real-time parameters, such as soil conditions, weather conditions, temperature, weed detection, crop diseases, and so on, in order to increase crop yield productivity. Also, AI helps farmers automate farming and helps to increase the crop yield with better quality using minimal recourses. AI-based technology helps farmers in various areas including water management, optimization of pest and weed management, soil health monitoring, crop selection, and crop monitoring.

REFERENCES

Ananthi, S., & Varthini, S. V. (2012). Detection and classification of plant leaf diseases. *International Journal of Research in Engineering & Applied Sciences, 2*(2), 763–773.
Breiman, L. (2001). Random forests. *Machine Learning, 45*(1), 5–32.
Dahikar, S. S., & Rode, S. V. (2014). Agricultural crop yield prediction using artificial neural network approach. *International Journal of Innovative Research in Electrical, Electronics, Instrumentation and Control Engineering, 2*(1), 683–686.
Drummond, S. T., Sudduth, K. A., Joshi, A., Birrel, S. J., & Kitchen, N. R. (2003). Statistical and neural methods for site-specific yield prediction. *Transactions of the ASAE 46*(1), 5–14.
Dwivedy, N. (2011). Challenges faced by the agriculture sector in developing countries with special reference to India. *International Journal of Rural Studies (IJRS), 18*(2), 1–6.
Fortin, J. G., Anctil, F., Parent, L., & Bolinder, M. A. (2011). Site specific early season potato yield forecast by neural network in Eastern Canada. *Precision Agriculture, 12*(6), 905–923.
Guruprasad, R. B., Saurav, K., & Randhawa, S. (2019). Machine learning methodologies for paddy yield estimation in India: A case study. *IGARSS 2019–2019 IEEE International Geoscience and Remote Sensing Symposium.*
Jaikla, R., Auephanwiriyakul, S., & Jintrawet, A. (2008). Rice yield prediction using a support vector regression method. *ECTI-CON 5th International Conference, 2,* 29–32.
Jain, N., Kumar, A., Garud, S., Pradhan, V., & Kulkarni, P. (2017). Crop selection method based various environmental factors using machine learning. *International Research Journal of Engineering and Technology, 4*(2).
Kumar, R., Singh, M., Kumar, P., & Singh, J. (2015). Crop selection method to maximize crop yield rate using machine learning techniques. *2015 International Conference on Smart Technologies and Management for Computing, Communication, Controls, Energy and Materials (ICSTM),* 138–145.
Liu, G., Yang, X., Ge, Y., & Miao, Y. (2006). An artificial neural network based expert system for fruit tree disease and insect pest diagnosis. *Proceedings of the International Conference on Networking, Sensing and Control, ICNSC'06,* 1076–1079.
Liu, J., Goering, C. E., & Tian, L. (2001). Neural network for setting target corn yields. *Transactions of the ASAE, 44*(3), 705–713.

Okori, W., & Obua, J. (2011). Machine learning classification technique for famine pre-diction *Proceedings of the World Congress on Engineering, 2*, 991–996.

Shah, V., & Prachi, S. (2019). Crop yield and rainfall estimation in Tumakuru District using machine learning. *International Journal of Scientific Research in Computer Science, Engineering and Information Technology*, 1094–1097.

Shastry, K. A., Sanjay, H., & Deshmukh, A. (2016). A parameter-based customized artificial neural network model for crop yield prediction. *Journal of Artificial Intelligence, 9*(1), 23–32.

Snehal, D., & Sandeep, V. (2014). Agricultural crop yield prediction using artificial neural network approach. *International Journal of Innovative Research in Electrical, Electronics, Instrumentation and Control Engineering, 2*, 683–686.

Sonka, S. (2014). Big data and the Ag sector: More than lots of numbers. *International Food and Agribusiness Management Association (IFAMA), 17*(1), 1–20.

Tripathi, M. K., & Maktedar, D. D. (2016). Recent machine learning based approaches for disease detection and classification of agricultural products. *International Conference on Computing Communication Control and Automation (ICCUBEA)*.

Wolfert, S., Ge, L., Verdouw, C., & Bogaardt, M.-J. (2017). Big data in smart farming – A review. *Agricultural Systems, 153*, 69–80.

2 Deep Learning Models for Object Detection in Self-Driven Cars

Anisha M. Lal and D. Aju

CONTENTS

2.1 INTRODUCTION

In the current era, autonomous vehicles have drawn much research interest for the manufacturing and production industry. The autonomous car is strictly self-directed if it has the ability to sense the environment and travel safely without any manual intervention. The self-directed car should be able to localize itself from the outside environment by keeping track of the objects moving around it. Self-directed cars take raw data from sensing devices, such as cameras, light detection and range (LiDAR), and RADAR, and then process that information in a form that is suitable

DOI: 10.1201/9781003038450-2

for further processing. The information captured from sensors is used to localize the car and track objects around the environment.

Self-directed cars mainly focus on two aspects: instantaneous localization and the mapping, tracking, and detection of objects. Here, tracking and detection of objects have become a more challenging task. Traditional methods are used extracting feature vectors from the image and then some Machine Learning models to recognize the object present in the image. Supervised Machine Learning algorithms require a more labeled dataset, which was a drawback. Then, the focus was on feature detection methods, such as SIFT and HOG features, which provided compromising results when combined with Machine Learning models. The current focus is on Deep Learning models, which are able to extract the features automatically and recognize the object better than traditional methods using Machine Learning. This chapter mainly focuses on the various Deep Learning models used for detecting objects in self-driven cars, which provide extraordinary results compared to the existing cutting-edge methods.

2.2   RELATED WORK

Deep Learning methodologies that are used for recognizing objects are far more superior and efficient. The authors explicate how Deep Learning methods (Fujiyoshi et al., 2019) are applied to object recognition in image recognition. Also, the recent trends in Deep Learning pertaining to self-governing vehicle driving are explored and explained. It is observed that in practical application, under the pretense of discernment, the outcome of the Deep Learning method is a serious challenge. Also, it is understood that it is appropriate and advisable to utilize natural language processing by integrating both the ocular and verbal elucidations for better and competent output.

A comprehensive study on cutting-edge Deep Learning techniques that are employed in self-governing vehicle driving (Grigorescu et al., 2020) is presented. The comprehensive study pertains to automatic vehicle driving architecture through AI (Artificial Intelligence), CNN (Convolutional Neural Network), RNN (Recurrent Neural Network), and the DRL (Deep Reinforcement Learning) model. This comparative analysis provides insight into the strengths and weaknesses of the existing Deep Learning methods, as well as the AI methodologies for automatic vehicle driving.

A proposal on an integrated real-time object detection approach (Naghavi et al., 2017) through the captured images in autonomous self-driving vehicles is presented. A new methodology has been modeled for object detection through a unified neural network that mainly foresees the bounding boxes and class likelihood. The proposed approach is superior to You Only Look Once (YOLO) in terms of performance boost-up of about 5.4% mAP.

A primary learning methodology for video-oriented object detection has been overviewed and subsequently investigated (Masmoudi et al., 2019) thoroughly with respect to the agile transportation system. Here, Support Vector Machine (SVM), Deep Learning Technique such as YOLO, and Single Shot Multi-box Detector (SSD) have been focused upon with respect to the autonomous vehicle system.

After comparing the above methods, it is noted that the YOLO and SSD performed with higher accuracy in object detection, whereas the SVM performed poorly in real-time object detection.

An algorithm that aids in identifying, cataloging, and tracking the classified objects with respect to the autonomous vehicle is presented. In the proposed method, YOLO is combined along with the data that is obtained through the laser scanner and is used to identify and catalog the objects and, subsequently, to calculate the different objects around the vehicle. Also, the ORB (Oriented Fast and Rotated Brief) feature set (Aryal, 2018) is fused and coupled with a navigational system (GPS/INS) through Kalman Filter and is used to match the similar objects from one image frame to another image frame. Henceforth, the results obtained help the vehicle navigate autonomously on roads by localizing the vehicle as well as the various objects surrounding them.

An approach for deep object detection without any expensive pre-processing or deep evaluations has been applied and examined with the help of the image bounding box predictions (Lewis, 2014). After all experimentations, it is noted that SimpleNet provided better accuracy in object prediction. The prediction accuracy rate is 11 fps with about 12.83% mAP accuracy.

TensorFlow along with MobileNet (Howal et al., 2019) has been leveraged to perform object detection for autonomous vehicle driving. Through the implementation of both TensorFlow and MobileNet, an efficacy of 85.18% is achieved in detecting the objects. Also, it is observed that the loss per step is 2.73, which monitors the reliability of the method being executed. A new hybrid Local Multiple System (LM-CNN-SVM) (Uçar et al., 2017) in accordance with the Convolutional Neural Networks (CNNs) and SVM has been proposed for object detection and pedestrian identification. CNN and SVM have been utilized due to their strong ability to extract the features as well as their robust cataloging quality, respectively. It is ascertained that the new proposed system achieved a higher accuracy value of 89.80% for 15 images and 92.80% for 30 images, respectively.

An empirical assessment for three different methodologies – SSD, R-CNN (Region-based CNN), and R-FCN (Region-based Fully Convolutional Networks) (Simhambhatla et al., 2019) – has been carried out to calculate and analyze how quick and precise they are in various aspects. The measurements of object detection such as pedestrians, traffic signals, vehicles, and other objects are performed on various driving conditions. The driving conditions that are considered for the experiment are day, night, rainy, and snowy. From the experiment, it is noted that the region-oriented methods have higher accuracy in all driving conditions.

An assumption of driving style (Basu et al., 2017) with the autonomous vehicle is explored and presented. It is assumed that aggressive drivers prefer aggressive vehicles and defensive drivers prefer defensive vehicles. Based on the vehicle drivers' driving preferences, experimentations are performed. It is observed that drivers tend to prefer the defensive mode of driving style, rather than that of their own.

A systematic testing tool, named DeepTest (Tian et al., 2018), has been designed, implemented, and evaluated to automatically detect inaccurate activities of the Deep Neural Network-driven vehicles. It synthesizes test cases leveraging different driving conditions in real-world scenarios, including different lighting conditions, blurring, fog, rain, and so on. The synthesized test cases are used to develop and test the DeepTest system for autonomous vehicles. The developed system helps to identify the potential erroneous behaviors that can lead to fatal accidents.

A novel method proposed to examine security attacks against the sign recognition system that can deceive autonomous cars called DARTS (Sitawarin et al., 2018) has been introduced. This very system has been strategized by creating two toxic signs. The effectiveness of the proposed attack, such as Out-of-Distribution attacks and In-Distribution attacks, has been evaluated and it is observed that the Out-of-Distribution outperformed the In-Distribution attacks on classifiers.

An argument on cars related to selfish cost optimization and courtesy (Sun et al., 2018) towards interactive drivers is projected here. The term courtesy is considered as an objective that measures the drivers' cost prompted by the drivers' behaviors. The absurdity of the human drivers can be measured by courtesy which can be incorporated into the autonomous car to plan accordingly. The effect of courtesy is analyzed and it is observed that in front of a human driver, the courteous automatic cars leave more space.

A coordinated set of Light Detection and Ranging sensors that provide an integrated view, much larger than the field of vision (Jayaweera et al., 2019), has been proposed. This subsequently confers the respective data to a centralized decision-making mechanism so that the desired actions are sent to the vehicles with a reduced bit rate and waiting time.

A detailed literature review pertaining to autonomous cars (Badue et al., 2020) that have the self-driving capability, prepared since the DARPA (Defense Advanced Research Projects Agency), notes that challenges are present. The surveyed autonomous cars come under the categorization of SAE (Society of Automotive Engineers) Level 3 or higher. Also, the research on the pertinent techniques for perceiving and decision-making has been reviewed. Finally, a detailed architecture of the self-driving cars and their depiction that is developed at the IARA (Intelligent Autonomous Robotics Automobile) has been presented.

A prototype was proposed with the help of Road Context Aware Intrusion Detection System (RAIDS), Neural Networks were executed, and, subsequently, various experiments on the Raspberry Pi (Jiang et al., 2019) board have been conducted with comprehensive datasets and useful and significant intrusion cases. Based on the experiments conducted, it is observed that the RAIDS outperforms the Intrusion Detection System significantly, except for the usage of the road context. Also, it is noted that they were able to attain an accuracy of 99.9% with lesser latency.

A useful study focusing on the personal characteristics, active driving experience, gender, and rhetoric elements (Zackova & Romportl, 2018) that aids in determining the presumptive preferences of the self-driving car's functionalities has been studied exhaustively. For the respective study and analysis, an online survey has been conducted with 430 respondents. It is also noted that in the ethical dilemma pertaining to self-driving cars, the third-person scheme is less prejudiced than the first-person scheme.

To estimate the efficacy of learned policies, a theoretical outcome cataloging equilibria (Lazar et al., 2019) on a parallel road network is established. Also, the experimental comparison between the learned policy outcomes and the superior feasible equilibria is performed. In mixed autonomous traffic, to reduce the bottleneck by swaying humans' routing decisions, deep reinforcement learning is employed for the first time.

2.3   SELF-DIRECTED CARS

Self-directed cars (SDC) play a major role in today's era to reduce accidents and to reduce the stress and tension released by the driver during driving. The main objective of the SDC is to drive automatically without human intervention. There are five major mechanisms involved in developing an SDC system. SDC mechanisms consist of the following:

2.3.1   COMPUTER VISION

Computer vision describes how the SDC sees the environment. Its main purpose is to find the objects near the car. This process is called image classification and it is mainly performed using Convolution Neural Network (CNN).

2.3.2   FUSION OF SENSOR DATA

Fusion of sensor data involves an analysis of the entire environment around the SDC. The SDC operates on data from different sensors and fuses them to provide useful information for the SDC to operate.

2.3.3   LOCALIZATION

Localization mainly involves the process of locating the position of SDC. Better driving decisions can be made based on the location of SDC in order to determine how to proceed further.

2.3.4   PATH PLANNING

Path planning deals with finding out the optimal route to proceed with the journey. This can be taken care of by algorithms, which provide optimal solutions with less distance and time.

2.3.5   CONTROL

Control is the process that is involved in steering the wheel. Now, based on all the information provided from the aforementioned stages, it is a challenging task to perform the task like a human rather than a robot.

Graphical recognition errands such as Image Cataloging, Object Localization, and Identification are the key features involved in autonomous vehicle applications.

CNNs are mainly used for the image classification task, to classify the image and identify the object present in the image. Object localization is mainly used to identify the location of the objects in a particular image. Bounding boxes aid in localizing various objects that exist in the image. Object identification is the process of blending image classification, as well as object localization. It is to recognize numerous objects and subsequently find the location of each and every object present in the image. In SDCs, object detection involves detecting cars, pedestrians, motorcycles, trucks, buses, trees, and other objects present in the frame. This chapter focuses on object detection models used for self-directed cars.

2.4 OBJECT DETECTION

Primarily, object detection in an image or video sequence is a computer-vision-based activity to locate multiple objects and recognize them. By drawing a bounding box around the desired objects, it locates them and highlights them in the image or video sequence. Machine Learning (ML)-based and Deep Learning (DL)-based approaches are the two broadly classified object detection methods. Conventionally, existing systems rely upon ML techniques that deal with extracting the features of an image and grouping pixels based on the object type. These extracted features from the image are then given as input into a regression model for further predicting the corresponding objects and their respective labels. Contrary to that, the cutting-edge DL methods focus on unsupervised learning of automatic feature extraction and classification of objects present in the image. Figure 2.1 shows the difference between the classification and localization of objects.

Object detection involves image classification and localization. Image classification using DL models is done using CNNs. CNN is used to recognize objects present in the image, by performing convolution operations on the images to classify objects. The main disadvantage of CNN is that it can detect or classify only a single object from an image. In order to overcome this disadvantage, the sliding window concept has been introduced.

Classification   Classification +
                               Localization

FIGURE 2.1 Classification and object detection.

2.5 REGION-BASED CONVOLUTIONAL NEURAL NETWORK (R-CNN)

To overcome the issue that occurred in CNN and identify multiple objects, the selective search approach with regions is extracted from the image and is called region proposals. Here, in lieu of focusing on all the regions present in the image, the selected regions can be focused on to find the desired outcome. The region proposals extracted from the image are created using a selective search algorithm, which works based on three major steps: First, it sub-segments the image into regions; second, it focuses on combining regions with similar properties using a greedy algorithm; and, finally, it generates regions in order to create region propositions.

The selected region propositions are wrapped and given as input to the CNN to generate a feature vector as its result. CNN is utilized as a feature extractor and the dense layer acts as output, consisting of all the extracted features. All these features having been extracted are used for further processing. The obtained features are then given as input to the SVM classifier to classify the objects present in the region proposals. The objects are detected in the region proposals along with the four offset values that increase the preciseness of the bounding boxes. Often, the extracted objects from the region proposals are not accurate to fit into the bounding box. To overcome the inaccuracy, the offset values are adjusted to fit in the bounding box.

The framework used for R-CNN is given in Figure 2.2. The main disadvantage of R-CNN is that it is quite difficult to implement in real-time because it generates many region proposals, and the computing features from each region proposal becomes a challenging task. The selective search algorithm used for generating region proposals is fixed and no learning happens in this stage, which leads to bad extraction of region proposals and that makes the whole process complex and time-consuming.

2.6 FAST REGION-BASED CONVOLUTIONAL NEURAL NETWORK (FAST R-CNN)

To overcome the disadvantage of R-CNN, which feeds region proposals to a CNN, a Fast R-CNN is introduced. It feeds the input image to the CNN and subsequently extracts the feature map. The region proposals are identified using a selective search algorithm from the feature map and then given as input into a Region of Interest (ROI) pooling layer. The fixed size ROI are then fed into a fully connected layer, and then, using the softmax layer, the objects are classified, and offset values are generated using bounding boxes.

Figure 2.3 shows the framework as well as the workflow for Fast R-CNN. Compared to R-CNN, Fast R-CNN provides better performance, in terms of quickness and accuracy. Once the input image is directly fed into the CNN, it automatically computes the feature map. Even though the region proposals are detected through a selective search algorithm, the respective detection process slows down in Fast R-CNN. Therefore, generating region proposals through a selective search algorithm has become a major defect in R-CNN and Fast R-CNN, and it, subsequently, affects the performance of both networks.

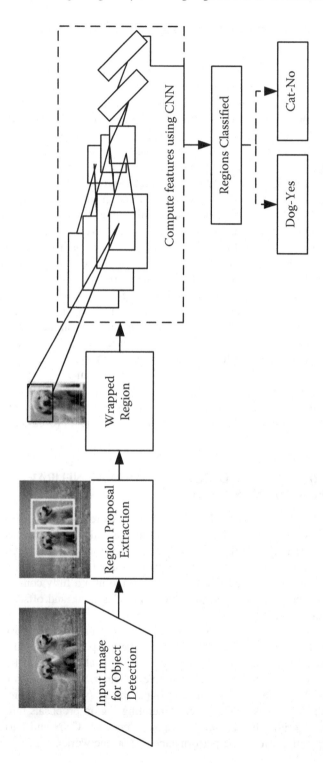

FIGURE 2.2 Framework for R-CNN.

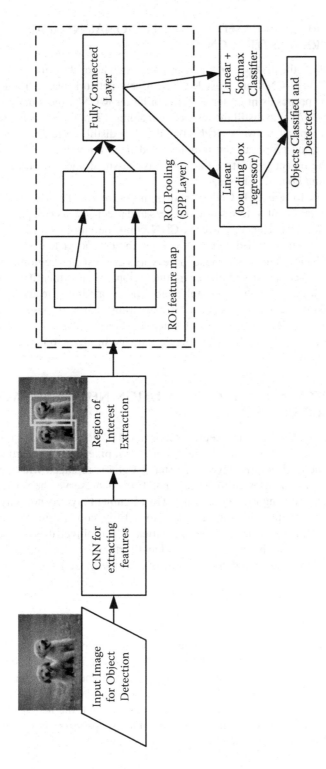

FIGURE 2.3 Framework for fast R-CNN.

2.7 FASTER REGION-BASED CONVOLUTIONAL NEURAL NETWORK (FASTER R-CNN)

In order to surmount the disadvantage of selective search algorithms that extract the region proposals in R-CNN and Fast R-CNN, a new network was introduced to extract the region proposals in Faster R-CNN. In Faster R-CNN, the input image is directly sent to the CNN to distill the respective features. From the extracted feature map, a Region Proposal Network (RPN) is utilized to distill the region proposals. The respective region proposals are remolded and fixed through the ROI pooling layer; further, the object present in the region is classified and the offset values are identified using bounding boxes.

The framework for Faster R-CNN is shown in Figure 2.4. The Faster R-CNN provides better attainment in terms of quickness and precision, as compared to R-CNN and Fast R-CNN. Hence, Faster R-CNN could be employed in detecting real-time objects, such as self-directed cars. Even though Faster R-CNN is faster in terms of speed and accuracy, because these works are based on region proposal extractions, there may be parts of objects in the regions. Hence, identifying every object on the image becomes complicated and sometimes leads to mis-classification. To overcome this, instead of extracting region proposals and not focusing on the complete image, object detection algorithms, such as YOLO, were introduced, which considers the complete image in order to detect objects present in the image.

2.8 MASK REGION-BASED CONVOLUTION NEURAL NETWORK (MASK R-CNN)

Mask Region-based Convolution Neural Network (Mask R-CNN) is the updated version of Faster R-CNN, which is mainly focused on pixel-level predictions. In Mask R-CNN, the ROI pooling layer of Faster R-CNN is replaced with the ROI alignment layer, which uses bilinear interpolation to preserve the spatial information extracted during the feature map. The alignment layer is not only used for boundary box prediction, it is also fed to a fully convolutional network to predict the mask, which means the semantics of the object detected. Mask R-CNN is used to locate objects at pixel level and further used for improving the precision of the objects detected. The framework for the Mask R-CNN is shown in Figure 2.5.

2.9 YOLO

YOLO is a real-time object identification technique. It is faster than Faster R-CNN because of its simple architecture. It is trained to do classification and bounding box prediction at the same time, unlike other R-CNNs. As the name suggests for YOLO, it looks only once for processing. It is sufficient that for predictions and classification, it requires only one forward propagation pass through the neural network. The object will be detected only once after a non-max suppression and the objects will be detected with the bounding box.

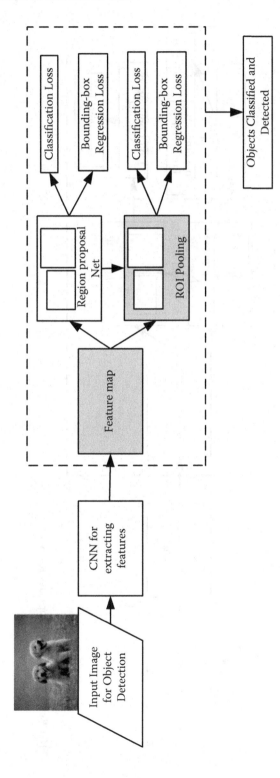

FIGURE 2.4 Framework for faster R-CNN.

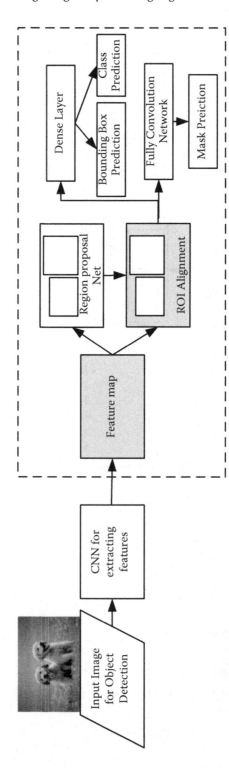

FIGURE 2.5 Framework for mask R-CNN.

The main advantage of YOLO is that it is much faster compared to other R-CNNs. It realizes the complete image during the training and testing phase and comes out with contextual information and is able to provide information about the objects and their appearances. For object detection, YOLO mainly focuses on learning generalizable images of objects so that when trained on natural images, this algorithm outperforms the other state-of-the-art algorithms. Nowadays, YOLO has come up with modifications and versions such as YOLO v1, YOLO v2, and YOLO v3.

2.10   YOLO V1 FOR SELF-DRIVEN CARS

YOLO v1 is a fast object detection algorithm in real-time, which is used for self-driven cars. In this chapter, YOLO v1 is specifically used with respect to self-driven cars. It is a challenging task to make cars drive on their own. For this, the car should be able to understand the environment and identify the objects around it. The best and efficient approach to instructing a car is to make it learn; there is a distinct kind of neural network that is called the Convolutional Neural Network (CNN). CNN has the amazing ability to understand spatial information even if it is rotated or skewed. This is done with the help of special layers called convolutional layers, which are used for extracting the features of the image. CNN is mainly used for classification and it has the ability to identify the class of the object. However, in self-driven cars, the main objective is to find out the class and also the location of the object. In addition to the location, there are numerous objects in the image with different classes that also have to be addressed. This is where object detection comes into the picture; the CNN can also be modified to locate the object using YOLO v1. YOLO v1 works on the basic principle behind CNN which is used for classification. YOLO v1 is incredibly fast as it utilizes 24 convolutional layers. Since it uses 24 layers, it can process up to 155 fps. This setup can be effortlessly executed in any self-driven cars.

In a self-driven car, by fixing the camera in the hood of the car, the images are captured at a fixed rate and stored in a database. The captured images are labeled by drawing a bounding box surrounding the object present. Here an example is given in Figure 2.6 showing the bounding box. The bounding box must be presented in an

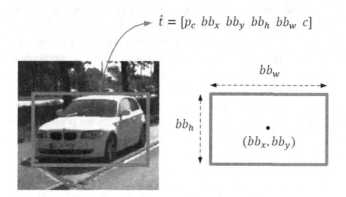

$$\hat{t} = [p_c \ bb_x \ bb_y \ bb_h \ bb_w \ c]$$

bb_w

bb_h

(bb_x, bb_y)

FIGURE 2.6   Bounding box.

appropriate form. Since this chapter focuses on self-driven cars, it targets six classes: "car", "person", "truck", "bike", "bus", and "cycle". Hence, the target variable of the bounding box is defined as given in Eq. (2.1).

$$\hat{t} = \left[p_c\, bb_x\, bb_y\, bb_h\, bb_w\, c_1\, c_2\, c_3\, c_4\, c_5\, c_6 \right]^T \tag{2.1}$$

where

p_c = probability/confidence of object present in the image.
bb_x, bb_y = coordinate value of the center of the bounding box.
bb_h = image height of the bounding box.
bb_w = image width of the bounding box.
c_1 = class of the object being detected, here c_1 = car.

Since bounding box coordinates are represented as x, y, x', and y', the target variable parameters can be written as:

$$bb_x = \frac{(x + x')}{2 * width\ of\ the\ original\ image}, \quad bb_y = \frac{(y + y')}{2 * height\ of\ the\ original\ image}$$

$$bb_w = \frac{(x|' - x)}{width\ of\ the\ original\ image}, \quad bb_h = \frac{(y|' - y)}{height\ of\ the\ original\ image}$$

The YOLO v1 employs a single deep CNN to predict the class probabilities and bounding boxes simultaneously. The YOLO v1 model is shown in Figure 2.7.

YOLO v1 performs a unified detection of bounding boxes. Instead of using a selective search algorithm for detecting bounding boxes or region proposals in R-CNN, YOLO splits the entire image into S X S grids and foretells the bounding box at each grid cell that contains sufficient information for localizing and detecting the objects present in the image. The grid cell is liable for identifying the object when the center of the object falls inside a particular grid cell. YOLO v1 executes a cataloging and localization problem on all the grids simultaneously and detects the object present in the grid. YOLO v1 can locate only one object in each grid, which is a limitation in this algorithm. If we consider S = 7, then YOLO can detect only 49 objects in the image. An object can be located in many grids, so, to solve this problem, non-max suppression is used.

Each cell predicts bounding boxes and for each box, a confidence score (C) is generated. This confidence score reflects how confidently the object is present in the model. Using this confidence score, the model is prevented from detecting the background. The confidence score will be 0 if no objects are present in the cell. If it is not 0, then the confidence score will be equivalent to the Intersection Over Union (IOU) among the target box, as well as the ground truth. Since the ground truth is created manually, it has 100% confidence. Any box with a high IOU will have high confidence in having the object in the target cell. The boxes with low confidence can be discarded. See Figure 2.8.

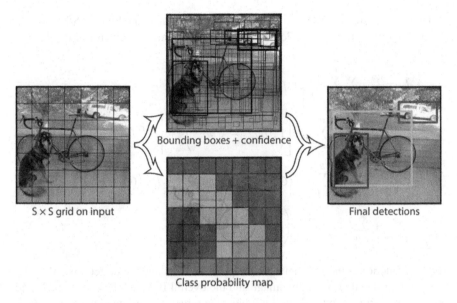

S × S grid on input

Bounding boxes + confidence

Final detections

Class probability map

FIGURE 2.7 YOLO v1 model for real-time object detection.

FIGURE 2.8 IOU using confidence score.

Here, since the S value is 7, there are 49 grid cells and each cell can predict 2 boxes, totaling 98 boxes. Most of the boxes will have a low confidence score and can be discarded.

In addition to the confidence score C, the model also generates four values bb_x, bb_y, bb_h, and bb_w to characterize the location and dimension of the predicted bounding box. The bounding box calculation is shown in Eq. (2.1) (Figure 2.6). Here, YOLO v1 is trained to identify six classes: "car", "person", "truck", "bike", "bus", and "cycle". Each grid cell computes six class probabilities that represent one grid cell for each class. Each and every single grid cell provides the option of selecting two bounding boxes in which the box with high confidence will be selected.

YOLO v1 uses CNNs to foresee numerous bounding boxes and class likelihood for that box. This respective network was inspired by the GoogleNet model for image cataloging. To obtain the final predicted outcome, this model utilizes 24 convolutional layers pursued by two fully connected layers.

The loss function in YOLO is calculated using Sum-squared Error (SSE), which is very easy to optimize. The loss function is represented as given in Eq. (2.2).

$$
\begin{aligned}
L_\theta = \lambda_{obj} \sum_{i=0}^{s^2} \sum_{j=0}^{B} l_{ij}^{obj} \left[(x_i - \hat{x}_i)^2 + \left| (y_i - \hat{y}_i) \right|^2 \right] \\
+ \lambda_{obj} \sum_{i=0}^{s^2} \sum_{j=0}^{B} l_{ij}^{obj} \left[\left(\sqrt{bb_{wi}} - \sqrt{\hat{bb}_{wi}} \right)^2 + \left| \left(\sqrt{bb_{hi}} - \sqrt{\hat{bb}_{hi}} \right) \right|^2 \right] \\
+ \lambda_{obj} \sum_{i=0}^{s^2} \sum_{j=0}^{B} l_{ij}^{obj} (C_i - \hat{C}_i)^2 + \lambda_{nobj} \sum_{i=0}^{s^2} \sum_{j=0}^{B} l_{ij}^{nobj} (C_i - \hat{C}_i)^2 \\
+ \sum_{i=0}^{s^2} l_{ij}^{obj} \sum_{c \in classes} \left(p_i(c) - \hat{p}_i(c) \right)^2
\end{aligned}
\tag{2.2}
$$

The loss function is the SSE between the predicted box and the target box. Also, the overall 49 grid cells exist in the image and the overall B boxes (B = 2) for each grid cell. The majority of the cells will not have the target box assigned to them. A kind of class-weighted classification is introduced with hyperparameters, λ_{obj} and $\lambda_{nobj,}$ which serve as a weighting factor for the loss value. The loss function is alienated into two localization losses: the confidence loss and the classification loss. The first localization loss is calculated with respect to the predicted box and the target box, which is shown in Eq. (2.3).

$$
\lambda_{obj} \sum_{i=0}^{s^2} \sum_{j=0}^{B} l_{ij}^{obj} \left[(x_i - \hat{x}_i)^2 + \left| (y_i - \hat{y}_i) \right|^2 \right]
\tag{2.3}
$$

The second localization loss foresees the square root of the bounding box width and height in lieu of calculating it directly, which is shown in Eq. (2.4).

$$
\lambda_{obj} \sum_{i=0}^{s^2} \sum_{j=0}^{B} l_{ij}^{obj} \left[\left(\sqrt{bb_{wi}} - \sqrt{\hat{bb}_{wi}} \right)^2 + \left| \left(\sqrt{bb_{hi}} - \sqrt{\hat{bb}_{hi}} \right) \right|^2 \right]
\tag{2.4}
$$

The confidence loss is reduced in case of there is no confidence score on the grids. Grids that do not contain the confidence score are pushed to zero; this may lead to diverging early. Hence, the confidence loss given in Eq. (2.5) is reduced.

$$
\lambda_{obj} \sum_{i=0}^{s^2} \sum_{j=0}^{B} l_{ij}^{obj} (C_i - \hat{C}_i)^2 + \lambda_{nobj} \sum_{i=0}^{s^2} \sum_{j=0}^{B} l_{ij}^{nobj} (C_i - \hat{C}_i)^2
\tag{2.5}
$$

The classification loss is used to total the errors of all class probabilities for the 49 grid cells present in the image. The classification loss is given in Eq. (2.6).

$$\sum_{i=0}^{s^2} l_{ij}^{obj} \sum_{c \in classes} \left(p_i(c) - \hat{p}_i(c) \right)^2 \tag{2.6}$$

Non-maximal suppression is used to interpret the problem of locating the same object in multiple grids. If a car is detected multiple times, then the following steps can be done to rectify this:

Step 1: Abandon all the boxes with less confidence score.

Step 2: Categorize the predictions from the box which has maximum confidence.

Step 3: Select the box having maximum confidence and the outcome is its prediction.

Step 4: Discard all the boxes with IOU less than the respective threshold with the box in Step (3).

Step 5: Replicate the process from Step (3) until the rest of the predictions are verified.

Limitations of YOLO v1: Each grid in the image can predict only two boxes and one class for each grid; this confines the number of neighboring objects predicted by YOLO v1. It can detect only 49 objects. YOLO v1 has a high degree of localization error and has low recall.

2.11 YOLO V2 FOR SELF-DRIVEN CARS

Version two of YOLO chiefly focuses on recall and localization error, preserving the classification accuracy. YOLO v2 is faster, better, and more advanced than the Faster R-CNN and the SSD. The updated and notable change of YOLO v2, compared to YOLO v1, is the normalization in the input layers, which is done by slightly altering and scaling the activations. Since batch normalization is added to the convolutional architecture, the mAP (mean Average Precision) has been improved by 2%, which reduces overfitting and the model has achieved regularity. The input size of the image is increased in YOLO v2, which indeed increases the mAP to 4% when training the YOLO v2 architecture DarkNet 19.

The most highlighted change is the introduction of anchor boxes. The classification and prediction are done in a single design in YOLO v2. The anchor boxes are primarily liable for predicting the bounding boxes, and k-means clustering is used for designing these anchor boxes for a given dataset. The main limitation in YOLO v1, that of identifying smaller objects, is attained in YOLO v2 by dividing the image into 13 × 13 grid cells; this provides a solution for identifying or localizing minor objects in the image and it is also efficient for identifying bigger objects. YOLO v2 can also be used to train with random images of different sizes, which improves the accuracy of the application on which it works.

YOLO v2 is designed using DarkNet 19 architecture, which has 19 convolutional layers that are used for extracting features, five max pooling layers, and a softmax layer that is mainly used for the classification of objects. Darknet is a neural network framework that is actually quick in object detection, which is a very significant feature in real-time for prediction in self-driven cars. YOLO v2 is better, faster, and stronger compared to other object detection models and has the ability to identify smaller objects; it also has the ability to classify objects with different dimensions and configurations. The YOLO v2 model is shown in Figure 2.9.

2.12 YOLO V3 FOR SELF-DRIVEN CARS

The main trade-off in object detection models is how accurately and quickly the objects are detected. YOLO v3 has an incremental improvement over the previous version, YOLO v2. YOLO v3 can detect objects accurately and classify objects in real-time applications. The incremental improvements compared to YOLO v2 are that bounding box predictions are done using logistic regression, and the objectiveness score is calculated for each bounding box. Instead of softmax used in the previous version to predict class probabilities, in YOLO v3, logistic classifiers are used for each class. Hence, multi-label classification is achieved.

YOLO v3 makes predictions similar to Feature Pyramid Networks (FPN), where, for each location in the input image, three predictions are made and features are extracted for each prediction. Each prediction contains bounding box, objectness, and class scores, which improve the ability of YOLO v3 in different scales. YOLO v3 is designed using DarkNet 53 architecture, which has 53 convolutional layers that are used as a feature detector. YOLO v3 is mainly suited for real-time applications that accurately classify and predict objects. Hence, it is well-suited for self-driven cars.

2.13 PERFORMANCE ANALYSIS

This chapter focuses on object detection models for self-driven cars. The performances of each model discussed in this chapter are evaluated based on its performance in classifying the objects, locating an object present in the image, and the precision with which it can locate objects in the image. The evaluations are done based on the models' predictions to the ground truth data. The models discussed in this chapter are pre-trained using Pascal VOC 2012 dataset and its performance is evaluated by means of mAP.

mAP determines the mean of all classes detected in the dataset used. mAP is based on the precision and recall of the classifier. In the object identification model, precision refers to the "false positive rate" and recall refers to the "false negative rate". The high precision and recall rate determines the classifier prediction and positively detecting all the objects present in the dataset. Average precision (AP) for a specific class is detected based on the precision-recall curve by varying the model threshold score, which predicts the true predictions of the object.

Figure 2.10 shows the performance of various object detection models with respect to mAP and run-time efficiency in frames per second (fps). The mAP is measured based on the intersection over the union value of 50 and 75. The first three

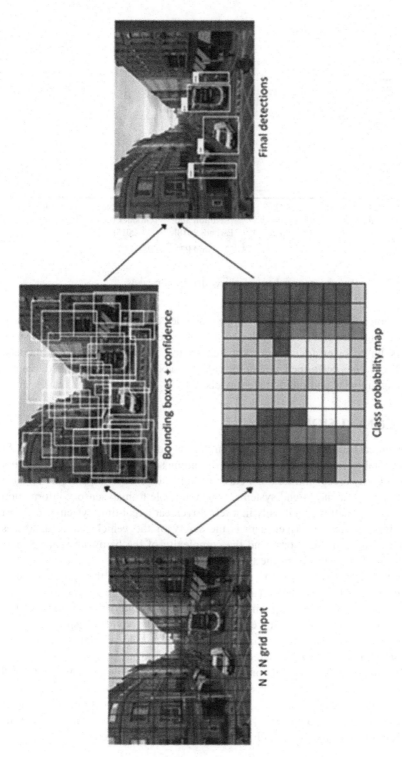

FIGURE 2.9 YOLO v2 object detection model.

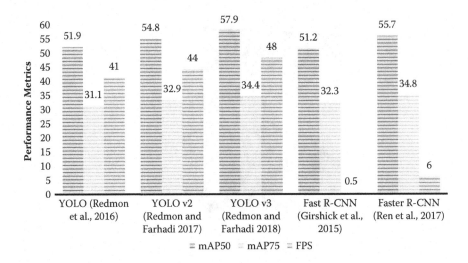

FIGURE 2.10   Performance analysis of object detection models.

models depicted in Figure 2.10 are single-stage detectors and the next two models are multi-stage detectors. Hence, due to the increased complexity, the multi-stage detectors have lower fps value compared to the single-stage detector models. The YOLO v3 has performed well in the case of object detection, compared to other single-stage models and multi-stage models.

2.14   CONCLUSION

Self-driven cars play a challenging role in the field of object detection. Current AI models are taken into consideration while designing self-driven cars. Deep Learning models not only focus on traditional classification and perception problems but also are widely used in end-to-end systems, which are able to map sensory information to steering information specifically in a self-driven car's real-time system. The most challenging task in self-driven cars is the safety of the vehicle; large databases are required to train the model and the complexity of the hardware requirements involved in the testing and in the training phase.

This chapter provides some of the Deep Learning models that can be used to locate objects present in the self-driven car application. The survey of some of the single-stage and multi-stage models has been done and their performance has been compared. In a single-stage model, YOLO v3 provides faster and more accurate results compared to the other models. Multi-stage models mainly focus on classification and they can detect the objects present in the input image. Faster R-CNN provides a better precision rate compared to other models. Mask R-CNN also establishes better results, when the main focus is on semantic segmentation of the object detected. Based on the aforementioned observation, this chapter provides insight for researchers whose focus is on object detection for real-time implementation in machines such as self-driven cars.

REFERENCES

Aryal, M. (2018). Object detection, classification, and tracking for autonomous vehicle, Master Thesis 912, Grand Valley State University.

Badue, C., Guidolini, R., Carneiro, R. V., Azevedo, P., Cardoso, V. B., Forechi, A., & de Paula Veronese, L. (2020). Self-driving cars: A survey. *Expert Systems with Applications*, *165*(1), 113816.

Basu, C., Yang, Q., Hungerman, D., Sinahal, M., & Draqan, A. D. (2017, March). Do you want your autonomous car to drive like you? In *2017 12th ACM/IEEE International Conference on Human-Robot Interaction (HRI)* (pp. 417–425), Vienna, Austria.

Fujiyoshi, H., Hirakawa, T., & Yamashita, T. (2019). Deep learning-based image recognition for autonomous driving. *IATSS Research*, *43*(4), 244–252.

Girshick, R. (2015). Fast R-CNN. In *Proceedings of the IEEE International Conference on Computer Vision* (pp. 1440–1448), Santiago, Chile.

Girshick, R., Donahue, J., Darrell, T., & Malik, J. (2014). Rich feature hierarchies for accurate object detection and semantic segmentation. In *Proceedings of the 2014 IEEE Conference on Computer Vision and Pattern Recognition (CVPR '14)* (pp. 580–587), Washington, DC.

Grigorescu, S., Trasnea, B., Cocias, T., & Macesanu, G. (2020). A survey of deep learning techniques for autonomous driving. *Journal of Field Robotics*, *37*(3), 362–386.

He, K., Gkioxari, G., Dollar, P., & Girshick, R. B. (2017). Mask R-CNN. In *2017 IEEE International Conference on Computer Vision (ICCV)* (pp. 2980–2988), Venice, Italy.

Howal, S., Jadhav, A., Arthshi, C., Nalavade, S., & Shinde, S. (2019, June). Object detection for autonomous vehicle using TensorFlow. In *International Conference on Intelligent Computing, Information and Control Systems* (pp. 86–93). Cham.

Jayaweera, N., Rajatheva, N., & Latva-aho, M. (2019, April). Autonomous driving without a burden: View from outside with elevated LiDAR. In *2019 IEEE 89th Vehicular Technology Conference (VTC2019-Spring)* (pp. 1–7), Kuala Lumpur, Malaysia.

Jiang, J., Wang, C., Chattopadhyay, S., & Zhang, W. (2019, December). Road context-aware intrusion detection system for autonomous cars. In *International Conference on Information and Communications Security* (pp. 124–142), Cham.

Lazar, D. A., Bıyık, E., Sadigh, D., & Pedarsani, R. (2019). Learning how to dynamically route autonomous vehicles on shared roads. *arXiv:1909.03664*.

Lewis, G. (2014). Object detection for autonomous vehicles. Stanford University.

Masmoudi, M., Ghazzai, H., Frikha, M., & Massoud, Y. (2019, September). Object detection learning techniques for autonomous vehicle applications. In *2019 IEEE International Conference of Vehicular Electronics and Safety (ICVES)* (pp. 1–5), Cairo, Egypt.

Naghavi, S. H., Avaznia, C., & Talebi, H. (2017, November). Integrated real-time object detection for self-driving vehicles. In *2017 10th Iranian Conference on Machine Vision and Image Processing (MVIP)* (pp. 154–158), Isfahan, Iran.

Redmon, J., Divvala, S., Girshick, R., & Farhadi, A. (2016). You only look once: Unified, real-time object detection. In *Proceedings of the IEEE Conference on Computer Vision and Pattern Recognition* (pp. 779–788), Las Vegas, Nevada.

Redmon, J., & Farhadi, A. (2017). YOLO9000: Better, faster, stronger. In *IEEE Conference on Computer Vision and Pattern Recognition (CVPR)*. Honolulu, Hawaii.

Redmon, J., & Farhadi, A. (2018). YOLOv3: An incremental improvement. arXiv preprint arXiv:*1804.02767*.

Ren, S., He, K., Girshick, R., & Sun, J. (2017). Faster R-CNN: Towards real-time object detection with region proposal networks. *IEEE Transactions on Pattern Analysis and Machine Intelligence*, *6*, 1137–1149.

Simhambhatla, R., Okiah, K., Kuchkula, S., & Slater, R. (2019). Self-driving cars: Evaluation of deep learning techniques for object detection in different driving conditions. *SMU Data Science Review*, *2*(1), 23.

Sitawarin, C., Bhagoji, A. N., Mosenia, A., Chiang, M., & Mittal, P. (2018). Darts: Deceiving autonomous cars with toxic signs. *arXiv preprint arXiv:1802.06430*.

Sun, L., Zhan, W., Tomizuka, M., & Dragan, A. D. (2018, October). Courteous autonomous cars. In *2018 IEEE/RSJ International Conference on Intelligent Robots and Systems (IROS)* (pp. 663–670), Madrid, Spain.

Tian, Y., Pei, K., Jana, S., & Ray, B. (2018, May). DeTptest: Automated testing of deep-neural-network-driven autonomous cars. In *Proceedings of the 40th International Conference on Software Engineering* (pp. 303–314), Gothenburg, Sweden.

Uçar, A., Demir, Y., & Güzeliş, C. (2017). Object recognition and detection with deep learning for autonomous driving applications. *Simulation*, *93*(9), 759–769.

Zackova, E., & Romportl, J. (2018). What might matter in autonomous cars adoption: First person versus third person scenarios. *arXiv preprint arXiv:1810.07460*.

3 Deep Learning for Analyzing the Data on Object Detection and Recognition

N. Anandh and S. Prabu

CONTENTS

DOI: 10.1201/9781003038450-3

3.1 INTRODUCTION

3.1.1 BASIC CONCEPT OF DEEP LEARNING

Deep Learning (DL) is a subset of machine learning. For the most part, DL has its base in Artificial Neural Networks (ANN), a computational model based on the working of the human brain. Like the brain of a human being, each of many computing cells, called "neurons", performs a fundamental procedure and connects to each other to make the decision. Deep Learning strategies are a class of machine learning algorithms that use nonlinear processing units at several layers to represent and transform.

Such DL is all about learning across many layers of a neural network correctly, efficiently, and without supervision. Also, it gained recent significance due to the facilitating developments in hardware computing. The preceding layer's output is used as input for each layer. In addition, various abstraction levels can obtain hierarchical representations. Such algorithms could be developed in a supervised or unsupervised manner and provide pattern analysis (unsupervised) as well as classification applications.

At present, three main factors have driven the rapid growth of Deep Learning: large data, efficient computational capacity, and new algorithms. To have a fundamental knowledge of Deep Learning, one should first explore the differences between descriptive and predictive analyses. Descriptive interpretation requires the concept of an intuitive mathematical model explaining the phenomena one wants to observe. This includes gathering process data, creating hypotheses on data trends, and validating such hypotheses, by comparing the descriptive models' effect that was created now to the actual results obtained (LeCun et al., 2015).

However, the development of such models is risky because there is often a possibility that scientists and engineers neglect to incorporate non-linear variables, either because they are ignorant or because of their inability to understand any complicated, hidden, or non-intuitive phenomena (O'Mahony et al., 2017). Such analysis of predictive type includes the exploration of rules of underlying phenomena and the development of a model of a predictive type that minimizes the errors in all potential intervening variables between the real and the expected outcomes (LeCun et al., 2015).

The traditional programming approach is ignored by machine learning, where a training framework replaces problem analysis, as such, a vast number of training patterns are fed into the system, which it learns and uses to measure new patterns (O'Mahony et al., 2019). There are several layers of Deep Learning models focused on neural networks, and each layer comprises several units. Every unit's activation y stands for input vector v's linear combination as well as learnable parameters w and base b, followed by a nonlinear function that is element-wise $f(\bullet)$, where a sigmoid function or small linear unit may be the function $f(\bullet)$. See Eq. (3.1). A Deep Learning network of layers, which are called architectures, with various communication structures.

$$y = f\,(wv + b) \qquad\qquad (3.1)$$

A variety of architectural setups of Deep Learning, such as Stacked Auto Encoders (SAEs), Convolutional Neural Networks (CNNs), Deep Belief Networks (DBNs), Recurrent Neural Networks (RNNs), and Generative Adversarial Networks (GANs), are made as a proposal and are used in a successful way in several areas. Such architectural setups are central as they may be expanded or combined for building a fresh framework for particular works.

3.1.2 Brief History of Deep Learning

Deep Learning is a modern technological feature that has arisen recently in the last few years. However, its historical development is longer, having been enriched over the last half a century. Such learning of deep type is dated as far back as the 1940s. It is in the areas of AI, as well as in the machine learning that comes from ANNs. Most Deep Learning architectures are built upon neural networks. The perceptron, suggested in 1957 (Rosenblatt, 1957), supervised learns a linear model that may be interpreted as $f(v,w) = wv + b$. Such mathematical modeling mimics the behavior of neurons in the human brain. This implies that it is also possible to consider Deep Learning as being comprised of many perceptrons as a generalized expression of linear rather than logistic regression, and it can imitate brain functions.

Since the 1980s, a few technological waves for ANNs, namely, Shallow Learning and Deep Learning, have improved the ways of learning methods. Each model input can be represented by several characteristics at this stage, and each function would activate a different neuron or hidden cell. The algorithm of backpropagation (BP) (Rumelhart et al., 1986) was made as a proposal in the mid-1980s for learning the artificial networks' parameters and later introduced a fresh upgrade to machine learning, based on a statistical model. In real-world implementations, the types of systems that are learning-based, namely, neural networks, have been seen as being more competitive compared to rule-based systems. These neural networks are otherwise known as Multi-Layer Perceptrons (MLP).

Research scholars gained some major strides in designing neural networks' sequences in the early 1990s. The primary emphasis of neural networks was on unsupervised learning. In order to generate a representation that has a low dimension but not labeled data, the models of neural nets were trained. As a consequence, since 1995, a number of model-based learning methods that were shallow have been made as a proposal, which has made them popular; these include Logistic Regression, Support Vector Machines, and Random Forests. It is possible to classify such shallow models as modeling with only one or no hidden layer. It is easy to train the shallow models and use them for solving sample problems with small sizes.

In contrast, networks of neural type require complex abilities to train and cannot be technically studied. Since 2000, when Internet strategies began evolving, such models of Shallow Learning became dominant in implementation in real-world scenarios, such as content scanning, advertising recommendations, and spam filtering methods. Due to the emergence of SVMs and the lack of backpropagation, the implementations in the earlier part of the 2000s were dominated by shallow models. However, in 2000, Aizenberg et al. first introduced the word

"Deep Learning" (Aizenberg et al., 2000). In 2006, when Hinton et al. published a paper in *Science* (Hinton & Salakhutdinov, 2006) – sponsored by the CIFAR (Canadian Institute for Advanced Research), Deep Learning started to attract interest. Neural Networks are used for learning knowledge representations that may be easily trained via pre-training done in a layer-wise way that may be carried out through unsupervised learning.

Deep Learning has had a steep rise in the world of academia since the breakthrough of 2006, and many universities are becoming the center of Deep Learning, such as universities of New York, Stanford, and Montreal, respectively. The U.S. government's DARPA (Defense Advanced Research Projects Agency) first funded such deep-type software learning in 2010. Google and Microsoft then used such learning technology to dramatically minimize speech recognition errors and have made the largest advance over the last 10 years in the area of speech recognition (Hinton et al., 2012). In 2012, for the first time, the aforementioned learning technology was used for identifying images in the ImageNet challenge and significantly increased output by 20%, contributing to the evolution of Deep Learning, as well as implementation in a variety of fields.

In other competitions, including the "International Conference on Document Analysis and Recognition" (ICDAR), a rigorous reading contest; Microsoft's "Common Objects in Context Challenge" (COCO); and the "Face Detection Dataset and Benchmark" (FDDB), the DL algorithms turned out to be the approach used by the mainstream for the last several years, surpassing the techniques of the shallow type, which focused on the feature being handcrafted. The emphasis for Deep Learning, since popular implementation, was initially on new and unsupervised learning strategies but has now been shifted to algorithms of a supervised learning type for using datasets of large-scale sizes.

3.1.3 ADVANTAGES OF DEEP LEARNING WITH TRADITIONAL LEARNING

Rapid advances in Deep Learning and system functionality have improved the efficiency and cost-effectiveness of vision-based applications, further speeding up their spread, including processing energy, memory space, energy usage, image sensor resolution, and optics. Deep Learning allows Computer Vision engineering teams to gain greater precision in activities such as image recognition, semantic segmentation, object identification, and Simultaneous Localization and Mapping (SLAM), as compared to conventional Computer Vision (CV) approaches.

Deep Learning also has superior versatility because, compared with CV architectures – which appear as additional domain-specific – models of CNN, as well as frameworks, may be re-trained using a custom dataset for any utilizing case. The typical method is to utilize a well-developed technique of CV for object identification, such as feature descriptors (SURF, SIFT, BRIEF, and so on). A phase known as the extraction of features was conducted for specific exercises, such as image recognition, before the advent of Deep Learning. The problem with this traditional approach is that in any given object, it is appropriate to select which features are relevant. If the number of groups for classification grows, the extraction of features becomes increasingly tedious.

Deep Learning implemented the idea of end-to-end learning, in which images' datasets that had gone through annotation with a class type of an object are available in every image and are only presented to the machine (Koehn, 1994). A model is then trained on the given data, where neural networks discover the underlying patterns in groups of images for each object and find out the most insightful and salient features for each particular type of object.

3.1.4 CONVOLUTIONAL NEURAL NETWORKS (CNNs)

In CV, the development of CNNs had a huge impact in the past few years. It caused a major change in the capability of recognizing objects (Voulodimos et al., 2018). Such escalation in progress is possible by the increased status of computing capacity. It is also due to the increase in the quantity of data availability for neural network training. SAEs and DBNs are fully connected networks since, between neighboring layers, both units are connected to each other. There is a vast range of parameters to be learned for certain types of data such as images, for networks that are fully connected. For avoiding such issues, the CNNs (LeCun et al., 1998) were suggested, influenced by the organization of the animal visual cortex, using the weight sharing technique to exploit structural units that are similar in different positions in a picture. This has dramatically reduced the number of parameters that are needed to be trained and makes the network equivalent in terms of input translations by sharing the convolutional weights in a local manner for the whole image. The simple CNN architecture shown in Figure 3.1 involves many different layers (Convolution, Pooling, and Fully Connected Layer) with different functions.

For detecting the feature or edges in a given image, CNN utilizes kernels (otherwise labeled as filters). A matrix containing values is termed as kernels that are learned for detecting certain characteristics, which are called weights. The key intention behind the CNNs is for transforming the kernel in a spatial way on an input image scan given, as their name implies, if the function it is intended to detect is available there or not.

A convolution process is done by calculating the kernel dot product as well as the input region where the overlapping of the kernel is done for providing a value reflecting how sure it is that a particular function is present. Then, the output of the

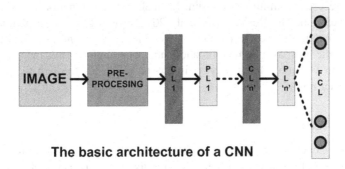

The basic architecture of a CNN

FIGURE 3.1   The basic architecture of a CNN.

convolution layer is summed up with a bias term. Now, feeding these to an activation function that is nonlinear to allow the learning of kernel weights. The function $\sigma(\cdot)$ is otherwise called an activation function. There can be several functions, such as the hyperbolic tangent tanh $[\sigma(x) = \tanh(x)]$, sigmoid $[\sigma(x) = (1 + e^{-x})^{-1}]$, and rectified linear units relu $[\sigma(x) = \max(0,x)]$.

The pooling layer typically leads the convolutional layer and it carries out nonlinear down-sampling. To enforce such down-sampling operations to pool layers, there are many nonlinear functions, of those the most common procedure is max pooling. In addition to the max pooling, other operations of a nonlinear type, such as pooling in average or L2 pooling of the norm type (Scherer et al., 2010), can also be done by the pooling layer. That layer of pooling is used in the architecture of CNN for steadily minimizing the spatial scale of intermediate representations, the parameters' count, and the computation amount, thereby controlling overfitting.

Subsequently, after many convolutional and pooling layers, the high-level logic in CNNs is applied with the layers that are fully connected (FC). Units in the FC layer provide complete access to all of the activations of the previous layer. As the weight count is decreased by convolution layer and pooling layers, the training of CNN may be carried out by an end-to-end approach by measuring the loss between the expected and true results. Typically, for various tasks, such as sigmoid cross-entropy loss, softmax loss, and Euclidean loss, separate loss functions are implemented.

LeNet (Kuo, 2016) was the first suggested CNN architecture in the early years, which formed the foundations for succeeding architectural setup. AlexNet (Krizhevsky et al., 2017) suggested, in 2012, expanding LeNet to larger, more convolution layers based on strong Graphics Processing Unit (GPUs). The ImageNet Large Scale Visual Recognition Challenge (ILSVRC) won such architectural setups and excelled at feature-based handcrafted approaches. Next, it was proposed to use a narrower filter and deeper frameworks in the popular VGG architecture (Chatfield et al., 2014). The network-in-network (Deng & Yu, 2014) was subsequently proposed for providing the brilliant and clear insight of using (1x1) convolution for providing the characteristics of convolutional layers with more combinational strength. The GoogLeNet, containing 22 layers, (Szegedy et al., 2015) was made as a proposal. It utilized the so-called "inception blocks" to get deeper layers. Such blocks may be considered as a form of network-in-network. It may also be used to minimize the dimensionality of the function maps by using (1x1) convolution. The ResNet (He et al., 2016) was suggested, in particular, for feeding the output of a few consecutive convolution layers and also for bypassing the input to the next layers. Much broader and deeper architectures may be made as a proposal in the future for achieving improved efficiency through the growth of computational power.

3.1.5 OBJECT DETECTION AND RECOGNITION

Considering Computer Vision, object detection deals with the detection in a visual image or video of instances of objects from a given class (Zhao et al., 2019). This specifies that the location object is displayed in the picture and scales one or more of the objects (Liu et al., 2020). Object detection is used to identify every object introduced to an image regardless of position, height, rendering, or other factors. Once the object is correctly identified, additional knowledge such as object class, object identification, and object detection is retrieved. Object detection consists primarily of two tasks: localization of objects and object classification. Object localization, by drawing a bounding box around it, specifies the position and size of one or more instances of an object. Object classification refers to the mechanism by which the label of a class is applied to the object. In order to detect, systems of object detection create a model from the training dataset, and, for generalization, a broad set of training data is required (Liu et al., 2017).

Looking at the broader family of Deep Learning architectural setups, a CNN, composed of a series of layers of a neural-type network, is used for CV and image processing (Pathak et al., 2018). It has a layer of input, a collection of hidden layers, and a layer of output. CNN takes the input as the image after processing it and then classifies it into a certain group. The use of CNN for object recognition has been applied for years, where arbitrary hidden layers' counts are utilized for detecting faces. The models of object detection execute the succeeding operations quickly: (a) collection of informative regions, (b) extraction of features, and (c) classification. The generic object detection pipeline, as shown in Figure 3.2, involves the training phase, the object, and the classification phase.

The traditional methods, such as the Viola-Jones detector (VJ), the feature-driven object detector, features of HOG using an SVM classification object detector, and Deep Learning approaches based on object detection (single-stage and two-stage methods) are several ways to identify the objects. The different approaches available for object detection and object detection applications are shown in Figure 3.3.

As there is a rise in the utilization of systems for video surveillance, face detection, vehicle tracking, and autonomous vehicle driving, the need for accuracy and fast object detection systems rose greatly. The performance of object detection usually involves a bounding box encircling an object with the confidence value calculated. Such detection could be a single-class object detection. Here, in a specific image, only one object is found. Yet, in multi-class object detection, it is important to recognize that more than one object is associated with different classes. In such detecting systems, they rely mostly on training examples as a large set because they are constructing a model to detect an object class.

3.2 DEEP LEARNING OBJECT DETECTION MODELS

The techniques may be separated into three key groups on the basis of the deep object detection pipeline shown in Figure 3.4: Anchor-Free Methods, Single-Stage (one-stage) methods, and Two-Stage methods. One-stage techniques extract and classify all the object proposals simultaneously. Two-stage approaches first produce proposals for a target object and then organize those proposals into particular categories. The two-stage methods possess a slower detection rate in a comparative sense and higher detection precision. However, single-stage techniques have faster detection speed and comparable precision for detection.

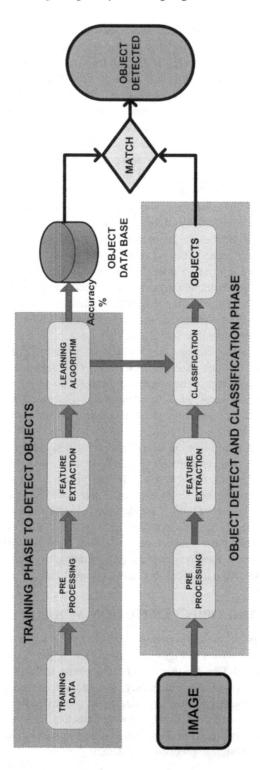

FIGURE 3.2 The generic object detection pipeline.

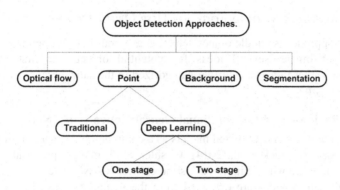

FIGURE 3.3 Different approaches and applications of object detection.

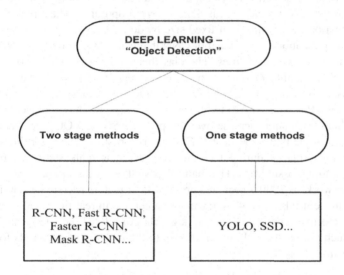

FIGURE 3.4 Different object detection methods in deep learning.

The R-CNN family (R-CNN, Fast R-CNN, Faster R-CNN, and Mask R-CNN) and SPPnet are primarily the regions' proposed approaches, some of which are associated with each other. You Only Look Once (YOLO) and Single-Shot Multi-Box Detector (SSD) are typically focused on regression and classification strategies. The anchors introduced in Faster R-CNN are bridged by the correlations of these two pipelines. Methods of two-stage models first locate the foreground objects and then categorizes them into a particular class; meanwhile, a single-stage detector skips the foreground object detection and takes uniform object samples via a grid. In the next part of this chapter, two-stage object detectors (R-CNN, Fast R-CNN, Faster R-CNN, and Mask R-CNN) and one-stage detectors (YOLO and SSD) are described, respectively.

3.2.1 Two-Stage Methodology for Deep Object Detection

Two-stage approaches handle object detection as a multi-stage operation. In view of the input image, several ideas for potential objects are first extracted. Subsequently, these ideas are further categorized in the same object categories by a trained classifier.

3.2.1.1 Region-based Convolutional Neural Network (R-CNN)

R-CNN is one of the most utilized models as a common object detector in the two-stage category (Girshick et al., 2014). Girshick et al. made a proposal of the R-CNN, which, as shown in Figure 3.5, is the earliest R-CNN detector.

Two thousand regions are extracted from the images by this process and are often referred to as region proposals. Such regions are CNN inputs, which generate as an output a 4096-dimensional feature vector. The architecture of R-CNN, using a search method of a selective type (Uijlings et al., 2013) where the extraction of the proposals of an object region set, is done. Every proposal of object region is made a transformation into an image of fixed size by doing a rescaling of it. Next, it is a question of application to the model of CNN that is pre-trained on ImageNet, that is AlexNet, used to extract features. The classifier of SVM (Dalal & Triggs, 2005) predicts the availability of objects inside every region proposal and it recognizes object classes by a bounding box. On the VOC-2007 dataset, RCNN increased mean Average Precision (mAP) to 58.5% from 33.7%. R-CNN achieved a mAP of 53.3%, which is a major increase over the prior PASCAL VOC 2012 challenge.

The major issue is the speed of R-CNN. It is too slow. The algorithm used for producing the proposal during the selective search is quite static. Therefore, it discards learning possibilities. The main disadvantage is that it takes more time to train the network, as 2000 object area proposals per picture need to be classified. As each test image needs about 47 s, it cannot be applied in real-time, and because the search of selective type approach is an algorithm of a fixed type, learning does not occur at such a rate. As such, it contributes to the creation of proposals for object regions of the bad type.

3.2.1.2 Fast Region-based Convolutional Neural Network (Fast R-CNN)

The detector of Fast R-CNN (Girshick, 2015) implementation was done by Girshick, which is an improved one of R-CNN. Such an architectural setup is

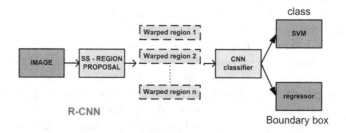

FIGURE 3.5 R-CNN – A region-based CNN detector.

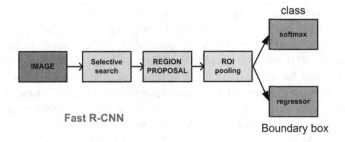

FIGURE 3.6 Fast R-CNN: A Faster Version of R-CNN.

depicted in Figure 3.6. It benefited from training both the detector and the bounding box regressor simultaneously. A feature map for the entire image was also computed by Fast R-CNN and features of the fixed-length area on the feature map were extracted. Fast R-CNN uses a region of interest (ROI) principle that has decreased time consumption. For extracting a feature map of fixed size that has a height and width that is fixed, the ROI pooling layer is used. It performs the maximum pooling procedure to convert features from the regions.

The classification layer had the responsibility to produce softmax probabilities over [C+1] classes (C classes with single background class), while four real-valued parameters were encoded by the regression layer for refining bounding boxes. The region classification, feature extraction, and bounding box regression steps in Fast R-CNN might perhaps entirely go through optimization end to end with no additional cache space for storing functionality. R-CNN extracts the CNN feature vector from every region proposal, while they are grouped by Fast R-CNN. Now, for one CNN only, the forward pass is produced. In the PASCAL VOC 2007 dataset, outcomes from the experiment inferred that Fast R-CNN could secure a 66.9% mAP score.

3.2.1.3 Faster Region-based Convolutional Neural Network (Faster R-CNN)

Despite the advancement in learning detectors, conventional approaches such as Edge Boxes or Selective Search, which were dependent on low-level visual signals and impossible to research in a data-driven manner, were still used in the proposal generation step. Faster R-CNN (Ren et al., 2015) was developed by Ren et al. to resolve this problem, as shown in Figure 3.7. A network, known as the Region

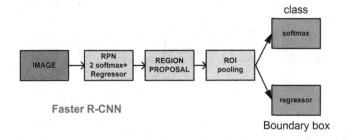

FIGURE 3.7 Faster R-CNN: a faster version of fast R-CNN.

Proposal Network (RPN), was implemented in Faster R-CNN shortly after the launch of Fast R-CNN to address the disadvantages of Fast R-CNN.

Faster R-CNN and Fast R-CNN are somewhat similar except in the case of RPN. Initially, ROI pooling is carried out first and then CNN and a few Fully Connected layers for the classifying of softmax and the bounding box regressor are fed to the pooled region. There is computational redundancy at the final level in spite of the fact that Faster R-CNN remained much quicker than Fast R-CNN. Further enhancements have been made to Faster R-CNN by Region-based Fully Convolutional Networks (RFCN) and Light Head R-CNN. The PASCAL VOC 2007 test set achieves a mAP score of 69.9%, which is a major increase for both detection and estimation performance over the Fast R-CNN.

3.2.1.4 Mask R-CNN

Mask R-CNN (He et al., 2017), which focuses on instance segmentation of an image, is expanded by Faster R-CNN and implemented by He et al., as shown in Figure 3.8. This is an extended feature of the Faster R-CNN and it also produces the object mask apart from the class label and a bounding box. Accurate identification is required in an instance segmentation activity. In addition, in Mask R-CNN, a mask branch for each ROI is used to predict that an entity operates in parallel with the regression divisions of the class mark and bounding box (BB). This generates three outputs: bounding box coordinates, a class label, and an object mask. An accurate detection task is necessary for instances of segmentation. The ROI align layer was implemented by Mask R-CNN, which maps the regions more specifically by fixing the position misalignment.

Mask R-CNN detects objects effectively in the input image or video. Then, it simultaneously produces a segmentation mask with higher quality for every instance of detected objects. Mask R-CNN achieves state-of-the-art efficiency of object recognition and segmentation of instances, based on multitask learning. This means that it will also help to enhance the efficiency of detection by joining the example segmentation task with the exercise of object detection. The only drawback is that it provides the network with small computational overheads and performs at a speed of nearly 5FPS.

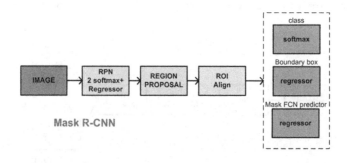

FIGURE 3.8 Instance segmentation using mask R-CNN.

3.2.2 One-Stage Methodology for Deep Object Detection

Unlike the multi-stage and two-stage technique processes, one-stage techniques are aiming at the simultaneous prediction of the object type and the position of the object. Compared to two-stage techniques, single-stage techniques possess a much higher speed of detection and a similar detection precision.

3.2.2.1 You Only Look Once – One-Stage Method

Redmon et al. have designed a real-time detector, known as YOLO (Redmon et al., 2016). This is the easiest, strongest, and shortest algorithm for object detection used in a real-time scenario. All previous algorithms for object detection use regions for identifying objects inside the image. However, the solution to YOLO is totally a varied one. The whole image is added to a CNN that is single. In Figure 3.9, the basic concept of YOLO is shown. It breaks the input image into an (SxS) grid and it is the responsibility of each grid cell to predict the entity centered in that grid cell. Bounding boxes "B" and their resulting confidence scores are predicted by each grid cell. The network of YOLO divides the entire picture into regions and predicts the boundary boxes and class probabilities for every region. Compared to Faster R-CNN, it is broadly understood as a unified network, it is quite fast, and it operates using a single CNN. The CNN used in YOLO was initially based on the GoogLeNet model. The revised version is called VGG-based DarkNet. At 45FPS, the network can process images in real-time, and 155FPS can be reached by the simplified version of Fast YOLO with better performance than most real-time detectors. The variants of YOLO are YOLOv1, YOLOv2, and YOLOv3. Here, YOLOv3 is the most recent variant. For object detection exercises in real-time, YOLO is a better and faster one. For precision improvisation, there is a possibility for training it end to end as it uses a CNN of a single type for estimation. YOLO runs at 155FPS (mAP = 52.7%). Its upgraded version runs at 45FPS (mAP = 63.4%) on the VOC-2007 dataset.

3.2.2.2 Single-Shot Multi-Box Detector (SSD)

Like YOLO, SSD (Liu et al., 2016) was also introduced by Liu et al., as shown in Figure 3.10, under the single-stage object detection model. In an era of Deep

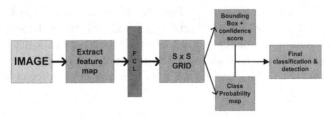

The basic architecture of a YOLO

FIGURE 3.9 The you only look once architecture.

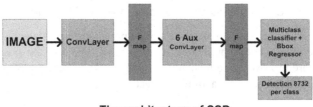

The architecture of SSD

FIGURE 3.10   Single-Shot Multi-Box detector architecture.

Learning, SSD is intended solely for real-time object detection. For detecting multiple objects inside the image, only one shot is needed. Localization of objects and recognition are all carried out in a single pass. SSD utilizes the prior trained VGG-16 (16-layer) model on the ImageNet dataset as its pivoted modeling to derive valuable image functionality. It implemented "multi-reference and multi-resolution detection" techniques to improve the SSD's detection accuracy, in particular small object detections.

SSD removed the RPN to increase the precision of Fast R-CNN's real-time speed detection. With Faster R-CNN, SSD reached comparable detection accuracy but appreciated the potential to do real-time inference. SSD-300 (mAP = 74.3%) achieves 59FPS on the VOC-2012 dataset, while SSD-500 (mAP = 76.9%) achieves 22FPS on the VOC-2007 dataset. The limitations of SSD are that with the number of default boundary boxes, precision improves at the expense of speed. Compared to R-CNN, the SSD detector has more grouping errors, but a minor localization error when working with related groups.

3.2.3   The Benchmark Deep Learning's Object Detection Models

The features and limiting aspects of the benchmark Deep Learning's object detection models, inclusive of the family of R-CNN, YOLO, and SSD, are presented in Table 3.1.

3.3   GENERAL DATASETS FOR OBJECT DETECTION

Since the models of object detection need an immense data quantity to be learned, the dataset plays a key role in the performance of these models. MS COCO (Lin et al., 2014) and PASCAL VOC (Everingham et al., 2015) are the simplified datasets available for object detection tasks. These datasets are in annotated form and are utilized to benchmark algorithms for Deep Learning. For object detection, MS-COCO and PASCAL-VOC datasets are often followed by all researchers.

3.3.1   Microsoft Common Objects in Context (MS-COCO)

MS-COCO is a dataset of 80 categories on a wide scale. There are three image sections in MS-COCO – with 118,287, 5000, and 40,670 images, respectively –

TABLE 3.1

Benchmark of Two-Stage Detectors and Single-Stage Detectors

Detector	Author & Year	Speed (FPS)	Features and Key Points	Limitations
R-CNN (Girshick et al., 2014)	Girshick et al. & 2014	<0.1	• To generate feasible objects using Selective Search • Perform image classification • Perform localization for multiple objects • Remarkable increase in efficiency over traditional models	• Extracted 2000 region proposals per image • Costly training in space and time • Unable to execute real-time implementation
Fast R-CNN (Girshick, 2015)	Girshick & 2015	<1	• Pooling layer of ROI • Proposed regions of fixed size • Implemented softmax for classification in favor of SVM • Disk storage is unnecessary for feature caching	• The slowest part in Fast R-CNN was Selective Search or Edge boxes • Selective search algorithms make it time-consuming to find a proposal • Struggle for real-time applications
Faster R-CNN (Ren et al., 2015)	Ren et al. & 2015	<5	• A region proposal network (RPN) has been implemented • Combine RPN and Fast R-CNN • Used for multiple image classification, detection, and segmentation	• May not support a plan for a Custom Region • To complete the entire process, several passes are required.
Mask R-CNN (He et al., 2017)	He et al., 2017	<5	• A quick, scalable, and powerful framework for segmentation of object instances • Utilized for instance segmentation • Introduced RoI-Align-layer	• Classification depends on segmentation • Training is expensive than other RCNN
YOLO (Redmon et al., 2016)	Redmon et al. & 2016	<25	• First efficient unified detector • Drop RP process completely • Single CNN for localizing and classifying • Used for immediate object detection • Generalizing for object demonstration	• Issue of objects' generalization with no-common aspect ratios • Combating the identification of small objects in a group • False localizations are the primary explanation for mistakes.
SSD (Liu et al., 2016)	Liu et al. & 2016	<60	• The first accurate and proficient integrated detector • Combine ideas from RPN and YOLO • Small CNN filters for predicting the category	• In detecting small objects, it is incapable. • Accurate, but much slower than YOLO, quicker than Faster R-CNN, but less accurate

TABLE 3.2

Summary of Deep Learning Object Detection Performances

S. No	Author & Model	"mAP" in % (MS-COCO)	"mAP" in % (PASCAL-VOC 2007)	"FPS"
1	Girshick et al. & RCNN (Girshick et al., 2014)	–	66	0.1 s
2	Girshick & Fast-RCNN (Girshick, 2015)	35.90	70.00	0.5 s
3	Ren et al. & Faster-RCNN (Ren et al., 2015)	36.20	73.20	6 s
4	He et al. & Mask-RCNN (He et al., 2017)	–	78.20	5 s
5	Redmon et al. & YOLO (Redmon et al., 2016)	–	63.40	45 s
6	Liu et al. & SSD (Liu et al., 2016)	31.20	76.80	8 s

validation, training, and testing. It has 330k images with 250k images labeled: 150 are object instances, 80 are categories of objects, 91 are categories of stuff, and 5 are captions per image. Features such as multi-objects per image, context recognition, and object segmentation are exhibited in this dataset. The MS-COCO test set's annotation information is not available.

3.3.2 PATTERN ANALYSIS, STATISTICAL MODELING, AND COMPUTATIONAL LEARNING (PASCAL) – VISUAL OBJECT CLASSES (VOC)

The PASCAL VOC is a mid-scale object detection dataset and contains only 20 classes of objects. The PASCAL VOC challenge offers structured image databases and a similar collection of tools for accessing accessible datasets for object type recognition activities and makes performance measurements and comparisons of different object detection methods from 2008 to 2012. In VOC-2007, there are three image sections: the training section (2501), the validation section (2510), and the testing of images (5011). In VOC-2012, there are three image splits: validation (5717), training (5823), and testing with images (10,991). The VOC-2012 test set's annotation information is unavailable.

A comparison of the results of different object detectors pivoted on Deep Learning on MS-COCO and PASCAL-VOC 2007 is shown in Table 3.2.

3.4 CONCLUSION AND FUTURE DIRECTIONS

Object detection refers to locating objects from digital images and classifying them. An issue of rudimentary nature for solving is object detection; existing methods have been developed. Also, in real-time implementations, there exists a wide potential to develop specific mechanisms and object detection as simple services. In a number of applications, CNN-pivoted object detectors are utilized with progressive outcomes from CNNs of the deep type of architectural setups. It is classified as either single-stage or two-stage target detecting model depending on methodology. As far as invariant occlusions, sizes, lighting, intra-class differences, and deformations are concerned, object detection in real-time scenarios using various GPU-pivoted embedded platforms was expected to be strong. For these reasons, the ability to identify small objects in a video is high, which decreases the efficiency of tracking systems for real-time objects.

A significant amount of data is needed for more accurate detection. To further enhance overall precision, training images are required with more object diversity. However, Scalable Proposal Generation Technique, Efficient Encoding of Contextual Knowledge, Detection pivoted on Auto Machine Learning, Evolving Object Detection Benchmarks, Low-shot Object Detection, Detection Task Backbone Architecture, and Incremental Learning are still several open challenges. To conclude, it is noted that the aforementioned considerations are not limited to positive future directions in this research area and research in such an area is yet to be extrapolated.

REFERENCES

Aizenberg, I. N., Aizenberg, N. N., & Vandewalle, J. (2000). *Multi-Valued and Universal Binary Neurons*.

Chatfield, K., Simonyan, K., Vedaldi, A., & Zisserman, A. (2014). Return of the devil in the details: Delving deep into convolutional nets. *arXiv preprint arXiv:1405.3531*.

Dalal, N., & Triggs, B. (2005). Histograms of oriented gradients for human detection. *2005 IEEE Computer Society Conference on Computer Vision and Pattern Recognition (CVPR'05)*, *1*, 886–893.

Deng, L., & Yu, D. (2014). Deep learning: Methods and applications. *Foundations and Trends in Signal Processing*, *7*(3–4), 197–387.

Everingham, M., Eslami, S. M. A., Van Gool, L., Williams, C. K. I., Winn, J., & Zisserman, A. (2015). The pascal visual object classes challenge: A retrospective. *Int. J. Comput. Vis.*, *111*(1), 8–136.

Girshick, R.B. (2015). Fast R-CNN. In *2015 IEEE International Conference on Computer Vision (ICCV)* (pp. 1440–1448).

Girshick, R., Donahue, J., Darrell T., J., Malik (2014). Rich feature hierarchies for accurate object detection and semantic segmentation. In *Proceedings of the IEEE Conference on Computer Vision and Pattern Recognition* (pp. 580–587).

He, K., Gkioxari, G., Dollár, P., & Girshick, R. (2017). Mask R-CNN. In *Proceedings of the IEEE International Conference on Computer Vision* (pp. 2961–2969).

He, K., Zhang X., Ren S., & Sun, J. (2016). Deep residual learning for image recognition. In *Proceedings of the IEEE Conference on Computer Vision and Pattern Recognition* (pp. 770–778).

Hinton, G. E., & Salakhutdinov, R. R. (2006). Reducing the dimensionality of data with neural networks. *Science*, *313*(5786), 504–507.

Hinton, G., Deng, L., Yu, D., Dahl, G. E., Kingsbury, B., & Sainath, T. N. (2012). Deep neural networks for acoustic modeling in speech recognition: The shared views of four research groups. *IEEE Signal Process. Mag.*, *29*(6), 82–97.

Koehn, P. (1994). Combining genetic algorithms and neural networks: The encoding problem.

Krizhevsky, A., Sutskever I., & Hinton, G. E. (2017). Imagenet classification with deep convolutional neural networks. *Commun. ACM, 60*(6), 84–90.

Kuo, C. C. J. (2016). Understanding convolutional neural networks with a mathematical model. *J. Vis. Commun. Image Represent., 41*, 406–413.

LeCun, Y., Bengio, Y., & Hinton, G. (2015). Deep learning. *Nature, 521*(7553), 436–444.

LeCun, Y., Bottou L., Bengio Y., & Haffner, P. (1998). Gradient-based learning applied to document recognition. *Proc. IEEE, 86*(11), 2278–2324.

Lin, T.-Y., et al. (2014). Microsoft COCO: Common objects in context. In *European Conference on Computer Vision* (pp. 740–755).

Liu, W., Wang Z., Liu X., Zeng N., Liu Y., & Alsaadi, F. E. (2017). A survey of deep neural network architectures and their applications. *Neurocomputing, 234*, 11–26.

Liu, W., et al. (2016). SSD: Single shot multibox detector. In *European Conference on Computer Vision* (pp. 21–37).

Liu, L., et al. (2020). Deep learning for generic object detection: A survey. *Int. J. Comput. Vis., 128*(2), 261–318.

O'Mahony, N., Murphy, T., Panduru K., Riordan D., & Walsh J. (2017). Improving controller performance in a powder blending process using predictive control. In *2017 28th Irish Signals and Systems Conference (ISSC)* (pp. 1–6).

O'Mahony, N., et al. (2019). Deep learning vs. traditional computer vision. In *Science and Information Conference* (pp. 128–144).

Pathak, A. R., Pandey, M., & Rautaray, S. (2018). Application of deep learning for object detection. *Procedia Comput. Sci., 132*, 1706–1717.

Redmon J., Divvala S., Girshick R., & Farhadi, A. (2016). You only look once: Unified, real-time object detection. In *Proceedings of the IEEE Conference on Computer Vision and Pattern Recognition* (pp. 779–788).

Ren, S., He, K., Girshick, R., & Sun, J. (2015). Faster R-CNN: Towards real-time object detection with region proposal networks. In *Advances in Neural Information Processing Systems* (pp. 91–99).

Rosenblatt, F. (1957). *The Perceptron, a Perceiving and Recognizing Automaton Project Para*. Cornell Aeronautical Laboratory.

Rumelhart, D. E., Hinton G. E., & Williams R. J. (1986). Learning representations by back-propagating errors. *Nature, 323*(6088), 533–536.

Scherer, D., Müller, A., & Behnke, S. (2010). Evaluation of pooling operations in convolutional architectures for object recognition. In *International Conference on Artificial Neural Networks* (pp. 92–101).

Szegedy, C., et al. (2015). Going deeper with convolutions. In *Proceedings of the IEEE Conference on Computer Vision and Pattern Recognition* (pp. 1–9).

Uijlings, J. R. R., Van De Sande, K. E. A., Gevers, T., & Smeulders, A. W. M. (2013). Selective search for object recognition. *Int. J. Comput. Vis., 104*(2), 154–171.

Voulodimos, A., Doulamis, N., Doulamis, A., & Protopapadakis, E. (2018). Deep learning for computer vision: A brief review. *Comput. Intell. Neurosci.*, 2018, Article ID 7068349, 13 pages.

Zhao, Z.-Q., Zheng, P., Xu, S., & Wu, X. (2019). Object detection with deep learning: A review. *IEEE Trans. Neural Networks Learn. Syst., 30*(11), 3212–3232.

4 Emerging Applications of Deep Learning

S. Karthi, M. Kalaiyarasi, P. Latha,
M. Parthiban, and P. Anbumani

CONTENTS

4.1 INTRODUCTION

4.1.1 MACHINE LEARNING

Machine Learning (ML) is a branch of Artificial Intelligence (AI) that utilizes programming to make accurate predictions without explicit modifications to

DOI: 10.1201/9781003038450-4

existing systems. ML is an investigative apparatus that computerizes that builds insightful models computationally. It can gain information, perceive patterns, and take decisions, without human intervention.

4.1.1.1 Machine Learning Types

Traditionally, ML is described as the process of an algorithm figuring out a way to become more precise. There are four essential steps in this process: supervised learning, unsupervised learning, semi-supervised learning, and reinforcement learning. The kind of algorithm that a data *aaaa* scientist may prefer to use depends on what kind of data they are dealing with (Figure 4.1).

- *Supervised Learning*: The researcher supplies algorithms with preparatory information and determines the variables the algorithms test for associations.
- *Unsupervised Learning*: Algorithms that train on unlabeled information are utilized in this type of ML. The algorithms filter any significant connection by collecting information. The information algorithms are instructed on both with expectations or plan make are foreordained.
- *Semi-Supervised Learning*: This approach to managing AI is used in a blend of the two past designs. Data scientists can deal with estimation named planning data, yet the model is permitted to examine the data in isolation and to develop its own educational assortment authority.
- *Reinforcement Learning:* Normally, fortress learning is used to show a PC to complete a multi-step measure for which undeniably described principles exist. To finish an errand, information researchers program a calculation and impart it certain or negative signs while it sorts out some way to complete an assignment. However, the algorithms generally decide on their own what steps to bring the way.

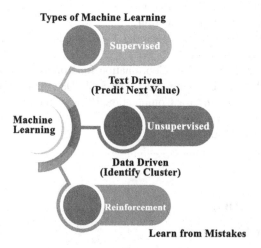

FIGURE 4.1 The basic types of machine learning.

4.1.1.2 How Machine Learning Overseen Works

Regulated ML permits the data scientist to set up the count for both the named inputs and the needed yields. Coordinated, for which the standards are portrayed unequivocally. Data analysts program a count to complete a task and give positive or negative movements toward it when working out how to finish a task. Nonetheless, the count chooses its own, what steps to bring the way. For the going with tasks, calculations for learning are reasonable:

- Binary Order. Partitioning information into two classes.
- Grouping of Multi-class. Choosing from multiple types of reactions
- Demonstration of Regression. Persistent characteristics foreseeing.
- Ensembling. To make an exact expectation, the forecasts of various AI models are joined.

4.2 SUMMARY OF DEEP LEARNING

Machine Learning uses algorithms for data parsing, learning from experience, and making decisions. Deep Learning structures complete computations in layers to make an "artificial neural network" that can make decisions.

Deep Learning – also called Deep Neural Learning or Deep Neuron Organization – is a part of ML in AI that can process unstructured or unlabeled information (Figure 4.2).

Deep Learning Algorithms has the following types:

- CNNs – Convolutional Neural Networks
- LTMNs – Long Short Term Memory Networks
- RNNs – Recurrent Neural Networks
- GANs – Generative Adversarial Networks

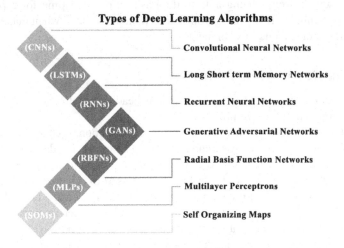

Types of Deep Learning Algorithms

(CNNs) —— Convolutional Neural Networks
(LSTMs) —— Long Short term Memory Networks
(RNNs) —— Recurrent Neural Networks
(GANs) —— Generative Adversarial Networks
(RBFNs) —— Radial Basis Function Networks
(MLPs) —— Multilayer Perceptrons
(SOMs) —— Self Organizing Maps

FIGURE 4.2 Types of deep learning algorithms.

- RBFNs – Radial Basis Function Networks
- MLPs – Multi-layer Perceptrons
- SOMs – Self Organizing Maps
- DBNs – Deep Belief Networks

Deep Learning has applications ranging from automatically deciding text headings to clinical diagnosis. Mechanized Driving: Deep Learning is used by car analysts to recognize objects naturally, for example, stop signs and traffic signals. Moreover, profound realizing, which limits wounds, is utilized to recognize people on foot.

4.2.1 Deep Learning Uses

Deep Learning is a technique of ML that urges PCs to do what humans do naturally, to learn as a viewable signal. A major development behind driverless vehicles is Deep Learning, which enables them to recognize a stop sign or to isolate a walker from a light post. In consumer electronics, including phones, tablets, TVs, and hands-free speakers, it is the way to achieve voice control. As of late, and taking everything into account, Deep Learning is getting a lot of consideration. In Deep Learning, a PC model sorts out some way to perform gathering tasks clearly from pictures, text, or sound. It achieves results that were ridiculous already. Deep Learning was first hypothesized during the 1980s, but there are two key reasons why it has just been valuable as of late:

1. Deep Learning requires tremendous amounts of data that are marked. Driverless vehicle creation, for instance, needs a great many pictures and a lot of video.
2. Deep Learning calls for a gigantic computational force. For Deep Learning, superior GPUs have equal engineering that is ground-breaking. This results in research groups being able to diminish preparation time for a profound undertaking in organization from weeks to hours, or less, when matched with bunches or a cloud environment.

4.2.2 The Deep Learning Development

1. In businesses going from self-sufficient heading to clinical gadgets, Deep Learning advances are utilized.
2. Autonomous Driving. In the quest for object detection, such as stop signs and traffic signals, vehicle researchers use Deep Learning. In addition, significant acknowledging, which cutoff points wounds, is used to recognize walkers.
3. Aerospace and Defense. Deep Learning is used to organize satellite things that discover areas of premium and to perceive troops ensured or risky zones.
4. Medical Research. Deep Learning is utilized by Cancer scientists to distinguish diseased cells naturally. UCLA research groups have made an inventive magnifying instrument that yields a high-dimensional informational

collection used to prepare a Deep Learning application to arrange malignancy cells precisely.

5. Industrial Automation. Deep Learning improves the well-being of laborers around large equipment by identifying when individuals or objects are inside an unprotected separation from machines.

6. Electronics: In programmed tuning in and discourse interpretation, Deep Learning is utilized. For example, Deep Learning applications are driven by home help gadgets that react to your voice and know your inclinations.

Neural association structures are used for most Deep Learning moves close, which is the explanation Deep Learning models are now and again insinuated as significant neural associations.

"Deep" normally implies the proportion of layers in the neural association that are hidden. Only two to three hidden layers are used in standard neural associations, while Deep associations may have as many as 150 layers.

By using gigantic game plans of checked data and neural association plans that gain incorporates direct from the data without the necessity for manual extraction of features, Deep Learning models are told.

4.2.3 Deep Learning Advantages

A Deep Neural Network (DNN) learns during the training process to discover useful patterns, such as sounds and images, in the digital representation of data. In fact, this is why we see more advances coming from Deep Learning in image recognition, machine translation, and natural language processing. Fast advancements in Deep Learning and enhancements in gadget usefulness, including processing limit, memory space, power use, picture sensor goal, and optics, have enhanced the plausibility and cost-adequacy of further quickening the extend of methods-based uses.

Contrasted with customary Computer Vision draws near, Deep Learning empowers Computer Vision specialists to accomplish more exactness in undertakings, including picture acknowledgment, semantic division, object location, and Simultaneous Localization and Mapping (SLAM). A new field of research in machine learning (ML) is Deep Learning. It requires different layers of artificial neural networks that are hidden. In wide datasets, the Deep Learning approach applies nonlinear transformations and high-level model abstractions. Recent advances in Deep Learning design have already made important contributions to AI in various ways.

This article provides a modern survey on Deep Learning contributions and novel applications. The following analysis chronologically presents how Deep Learning algorithms have been used and in what main applications. In addition, in typical applications, the superior and advantageous Deep Learning methodology and its hierarchy in layers and nonlinear operations are presented and contrasted with the more traditional algorithms. In addition, the state-of-the-art survey offers a general overview of the new concept and the ever-increasing benefits and popularity of Deep Learning.

4.3   DEEP LEARNING APPLICATIONS IN RECENT FIELDS

4.3.1   FRAUD DETECTION

Deep Learning the hang of: arising patterns, applications and study issues Mu-Yen Chen1 • Hsiu-Sen Chiang1 • Edwin Lughofer2 • Erol Egrioglu3,4 Online distributed: 11 April 2020 Springer-Verlag GmbH Germany, part of Springer Nature 2020. The extraordinary issue is planned to unite different accomplishments in innovative work for the investigation of strategies, applications, and difficulties confronting the development of workmanship. As follows, a short synopsis of the articles is introduced and discussed.

In this extraordinary issue, the main topic centers on "Parameter decrease in Deep Learning models." To assemble the stock value forecast model, Cao and Wang (2019) actualized the key part examination (PCA) and back engendering (BP) neural organization calculation. The second subject in this extraordinary issue centers around "Enhanced strategies for translation and thinking in Deep Learning models." An investigation on utilizing Deep Learning for peculiarity discovery was distributed by the exploration group at Cloudera Fast Forward.

They are creating Applied Machine Learning Prototypes as a component of Cloudera's progressing item improvement endeavors, which will send a total example AI project in your CML/CDSW model. The second Applied Machine Learning Prototype made accessible is for the improvement of a model for recognizing misrepresentation.

These are models that will assist you with making a working AI model in CML. The models contain information from the source and stroll through various advances:

- Discover the data assortment
- Progress a policy to generate a typical
- Train the prototypical to
- Organize the prototypical
- Generate and organize an application

The approach surveyed in this original includes numerous phases need to survey (See Figure 4.3):

1. Data Intake. Transfer the un-processing information to a more appropriate loading site.
2. Investigation of Data. Develop a strategy to shape the prototype.
3. Prototype. Training-train the plan-based prototype.
4. Distribution of the Prototype. Bringing the model live and into development.
5. Distribution of framework. Deploying a use that interrelates with the prototype.

4.3.2   AUTONOMOUS CARS

To impart the roads to human photographs, roused by propels in Deep Learning. None of them, in any case, consolidate the force of Deep Learning-based locators

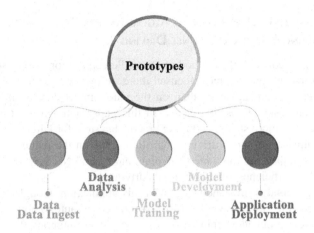

FIGURE 4.3 The prototype of fraud detection.

with past guides to distinguish the condition of the traffic signals drivers, independent earthly vehicles should be fit for seeing traffic signals and understanding their present statuses. Human drivers can rapidly find the proper traffic lights more often than not. A typical methodology for self-driving vehicles is to join distinguishing proof with earlier guides to manage this issue. Be that as it may, for the recognizable proof and acknowledgment of traffic signals, extra arrangements are required. Deep Learning techniques, including traffic-related issues, have shown incredible execution and speculation capacity. Some new works have utilized some cutting-edge profound indicators to find (and further perceive) traffic signals from 2D camera concerned. See Figure 4.4.

On this premise, this investigation proposes to consolidate the force of Deep Learning-based recognition with the past guides utilized by our IARA (Intelligent Autonomous Robotics Automobile) car stage to perceive the connected traffic signals of predefined courses. The cycle is isolated into two stages: a disconnected stage for the development of guides and comment of traffic signals; and an online stage for discovery and recognizable proof of the connected traffic signals. Five experiments (courses) in the city of Vitória were tried for the proposed plot, each case comprising of a video succession and an earlier guide with the suitable traffic signals for the street. The outcomes indicated that the proposed approach is prepared to accurately distinguishing the necessary traffic signal along the street (Rumelhart et al., 1986).

FIGURE 4.4 Autonomous cars working process.

4.3.3 GoogleNet Deep Learning Algorithm for Autonomous Driving Using GoogleNet Driving

In this paper, we consider the method of Direct Perception for autonomous driving. Previous efforts in this area have focused more on the extraction of road markers and other vehicles on the scene than on the autonomous driving algorithm and, under practical assumptions, its results. In this paper, our key contribution is to implement a new, more stable, and more practical system for Direct Perception and a corresponding autonomous driving algorithm. First, in the feature traction competitions, to compare the top three Convolutional Neural Networks (CNN) models and assess their efficiency for autonomous driving.

The experimental results showed that in this application, GoogLeNet performs the best. Therefore, we suggest a Deep Learning-based autonomous driving algorithm and refer to our algorithm as GoogLenet for autonomous driving (GLAD) (Aizenberg et al., 2000). GLAD makes no unrealistic claims about the autonomous vehicle or its environment, unlike previous efforts, and uses only five parameters of affordance to monitor the vehicle as opposed to the 14 parameters used by previous efforts. The results of our simulation show that the proposed GLAD algorithm exceeds previous Direct Perception algorithms on empty roads and when driving with other vehicles around it. See Figure 4.5.

4.3.4 Deep Learning to Self-Driving Car: Chances and Challenges

The human world is being revolutionized by Artificial Intelligence (AI). In the automotive industry, researchers and engineers are aggressively pushing autonomous driving methods focused on Deep Learning. However, it must first undergo a strict evaluation of functional protection before a neural network makes its way into series production vehicles. The opportunities and difficulties of implementing Deep Learning for self-driving cars (Hinton & Salakhutdinov, 2006).

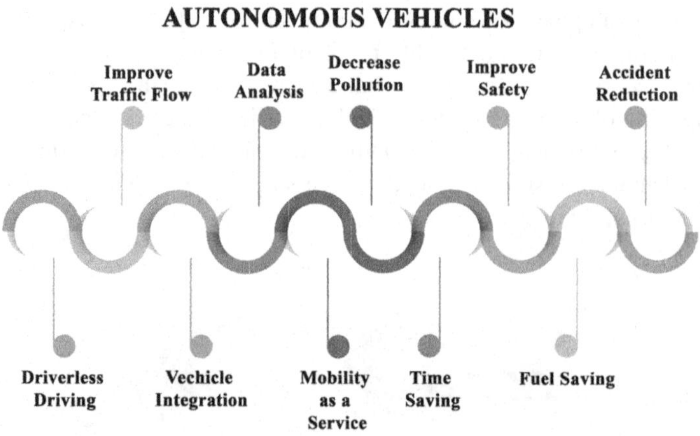

FIGURE 4.5 The basic architecture of a CNN.

4.3.5 DEEP LEARNING GROKKING

Deep Learning technology, a form of AI, teaches computer systems to analyze through the use of neural networks, a human brain-stimulated technology. Thanks to Deep Learning, on-line textual content translation, self-driving vehicles, customized product reviews, and automatic voice assistants are just a few of the thrilling new developments possible. The book *Grokking Deep Learning* teaches you to create neural networks in Deep Learning from scratch! Seasoned Deep Learning professional, Andrew Trask, demonstrates for you the technological know-how below the hood in his interesting style, so you grok each element of schooling neural networks for yourself.

You can teach your very own neural networks to look at and recognize photographs using simply Python and its math-assisting library, NumPy; translate textual content into exclusive languages; or even write like Shakespeare! You'll be completely organized to transport directly to learning Deep Learning structures while you are finished. What's interior The technological know-how in the back of Deep Learning Building and schooling your very own neural networks Privacy concepts, consisting of federated Learning Tips for persevering with your Deep Learning quest About the Reader For math and intermediate programming abilities at excessive faculty stage readers. About the author, Andrew Trask is an Oxford University Ph.D. student and a DeepMind studies scientist. Formerly, Andrew was a Digital Reasoning's researcher and analytics product manager, in which he became knowledgeable about the world's biggest artificial neural community and helped lead the Synthesys cognitive computing platform analytics roadmap (Hinton et al., 2012).

4.3.6 ENABLING IMMERSIVE SUPERCOMPUTING AT JSC, LESSONS LEARNED

Convincing eventualities for immersive supercomputing are have a look at and evaluation of massive volumes of records from experimental simulations, in-situ visualization and application management. The Jupyter (or JupyterLab) open-supply software program is a technique that has already been used efficiently in a variety of clinical disciplines. Jupyter is ideal for the incorporation of various kinds of plans and program design strategies in a single boundary, through its obvious and elastic web built architecture. For clinical programs at supercomputer centers, Jupyter's multi-person functionality through JuypterHub excels. See Figure 4.6.

It conveys the client neighborhood workspace and the relating workspace at the HPC frameworks. A smooth and direct web get admission to for starting and interfacing with login or figuring hubs with Jupyter or JupyterLab on the Jülich Supercomputing Center (JSC) is gotten a work to meet the necessities for more prominent intuitiveness in supercomputing and to open up new chances in HPC. The motivation, usage, particulars, and requesting circumstances of allowing intelligent supercomputing, notwithstanding wants and plausible fate work, can be routed to verify the adaptability of the fresh out of the plastic new cycle.

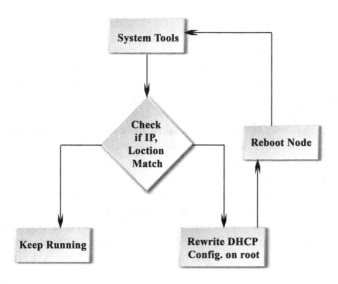

FIGURE 4.6 Enabling super computer.

4.3.7 WIDE-RANGE DEEP LEARNING ON BIG SCALE

Deep Learning is created from the concept of connectionism: at the same time as an character organic neuron or an character characteristic in a version of device learning isn't always sensible, sensible conduct may be tested with the aid of using a big populace of those neurons or functions running together. The reality that the wide variety of neurons need to be excessive is clearly vital to emphasize. The speedy upward push in the length of the networks is one of the fundamental elements answerable for enhancing the accuracy of the neural network and enhancing the issue of obligations they could remedy among the 1980s and today.

4.3.8 FAST CPU IMPLEMENTATION

The CPU of a single computer has traditionally been used to educate neural networks. Today, this approach is usually deemed inadequate. Now, we basically use GPU computation or the CPUs of numerous interacted computer systems organized together. Previously transferring to those high-priced setups, researchers labored with difficulty to expose that CPUs could not cope with the excessive computational workload desired with the aid of using neural networks. A examine of a way to put into effect powerful numerical CPU code is past this book's scope, however right here emphasize that for particular CPU families, cautious implementation can yield sizable improvements.

In 2011, for example, the fine CPUs to be had may want to run neural community workloads faster by the use of constant-factor mathematics instead of floating-factor mathematics. By designing a cautiously tuned constant-factor implementation, Vanhoucke et al. (2011) acquired a threefold acceleration and completed a stable variable factor structure. Every new CPU version has excellent overall performance features, yet, variable feature implementations can be quicker, as well. The essential perception is that by diligent specialism of arithmetical calculation routines, it could produce a sizable payout. Additional approaches contain optimizing facts systems to keep away from cache misses, with the aid of using the use of vector instructions further to find out whether or not to apply constant or floating points. Many device mastering researchers push aside those implementation details; however, the version's accuracy suffers whilst the overall performance of an execution limits the version's scale.

4.3.9 Large-Scale Implementations Distributed

In positive instances, the computing sources on a single computer are insufficient. Therefore, the education and inference workload needs to be done with numerous computers. It is easy to spread inference, for the reason that a separate computer can run every instance needed to process. This is stated as. Parallelism of facts Model parallelism also can be achieved, while many machines perform collectively on a unmarried facts point, with every system walking a specific fragment of the prototypical. This is possible for each guidance and interpretation. Information correspondence is much more difficult at some point of guidance. For a unmarried SGD phase, can boom the dimensions of the minibatch used, however in phrases of optimization efficiency, normally get much less than linear returns. Allowing numerous machines to measure a variety of steps of gradient descent in parallel might be easier.

4.3.10 Speech Recognition

The function of speech reputation is to map into the corresponding collection of phrases supposed through the speaker an acoustic sign containing a spoken herbal language utterance. Let $X_1 = (x_1(1), x_1(2), ..., x_1() T_1)$ represent the audio enter vector collection (historically produced through excruciating the acoustic into 20 ms edges). Maximum speech reputation structures use specialized hand-designed functions to preprocess the data, however, a few Deep Learning structures research functions from uncooked enter from Jaitly and Hinton (2011). Let $y = (y1, y2, ..., yN)$ denote the collection of goal production (commonly a chain of phrases or characters).

The function of automatic speech recognition (ASR) includes building an f-ASR characteristic that, given the acoustic collection, computes the maximum in all likelihood linguistic collection:

$$y X_1 f_1 * ASR() = \text{argmax} X_1 y P *(=) y X_1$$

Where $P*$ is the true conditional distribution relating the inputs X to the targets y.

Natural Language Processing: Natural Language Processing (NLP) is the use of human dialects, along with English or French, with the guide of utilizing a PC. PC applications typically consider and emanate specific dialects intended to allow efficient and unambiguous parsing with the guide of utilizing simple applications. All the more clearly happening dialects are routinely equivocal and resist formal depiction. Characteristic language preparation is comprised of projects along with framework interpretation, wherein the student needs to contemplate a sentence in a solitary human language and radiate an equivalent sentence in some other human language.

Numerous NLP programs are founded on absolute language molds that define a chance circulation over successions of expressions, characters, or bytes in a home-grown language. Similarly, as with the elective projects referenced on this part, extremely acknowledged neural local area methodologies might be proficiently done to natural language preparing. Be that as it may, to get good in general execution and to scale well for large projects, a couple of space-specific proce-dures arise as significant. To develop an efficient variant of the home-grown language, one needs to typically utilize procedures that may be particular for handling consecutive information. Much of the time, select to regard home grown language as a chain of expressions, in inclination to a chain of individual char-acters or bytes. Since the entire scope of suitable expressions is so monstrous, word-fundamentally based absolutely language styles need to perform on an ex-tremely high-dimensional and meager discrete zone. A few methods had been progressed to make designs of this kind of zone efficient, each in a computational and from a measurable perspective.

In language models, device conversion, and natural language processing, Deep Learning techniques have been very successful because of using embedding for signs (,) and phrases (,) (Rumelhart et al., 1986; Deerwester, 1990; Bengio et al., 2001). Such embedding replicates the semantic comprehension of human phrases and ideas. The advent of embedding for terms and for relationships between phrases and data is a studies frontier. Additionally, search engines use machine learning for this cause, however, tons extra stays to be carried out to enhance those extra state-of-the-art representations.

4.3.11   LEARNING HAND-EYE COORDINATION FOR ROBOTIC GRASPING WITH DEEP LEARNING AND LARGE-SCALE DATA COLLECTION

For robot greedy from monocular pictures, outline a studying primarily based to-tally method to hand-eye coordination. The skilled a extensive convolutionary neural community to are expecting the probability that task-area movement of the gripper could bring about powerful greedy to research hand-eye coordination for greedy, the use of simplest monocular digital digicam photos unbiased of digital digicam calibration or the contemporary robotic pose. This permits the research community to examine the gripper's spatial courting with gadgets within the scene, consequently studying hand-eye coordination.

Then use this community to servo the gripper in actual time to reap a hit grasps. This machine identifies large-scale experiments that were performed on two large robot structures. In the primary test, over several months, approximately 800,000 hold close tries have been obtained, the use of among 6 and 14 robot manipulators at any given time, with versions in digital digicam positioning and put on and tear of the gripper. In the second test, to collect a dataset that includes over 900,000 comprehension tries, they used a specific robot platform and eight robots. To determine the transition among robots, the second robot platform changed into used and the diploma to which information from a specific organization of robots may be used to assist studying. The effects display that our approach achieves green actual-time control, can seize novel artifacts effectively, and via way of means of non-stop serving corrects errors. Our switch test additionally illustrates that it is far more viable to combine information from numerous robots than to research extra accurate and efficient grasping.

4.3.12 DEEP LEARNING FOR COMPUTATIONAL CHEMISTRY

In computational chemistry, the upward thrust and fall of synthetic neural networks are recorded. Yet, after a long time, now we are seeing a revival of interest in Deep Learning, a multilayer neural network primarily based totally gadget studying algorithm. In this analysis, we offer an introductory assessment of the concept and precise uses of DNNs that differ from conventional computational methods used in chem-informatics. In this survey, a basic appraisal into the idea of DNNs and their specific homes that recognize them from traditional contraption contemplating calculations used in cheminformatics. See Figure 4.7.

This part represents its universality and wide appropriateness to an immense assortment of requesting circumstances in the field, comprehensive of quantitative construction movement relationship, computerized screening, protein shape forecast, quantum science, substances plan, and possessions expectation, with the guide of utilizing giving a high-level perspective on such a rising projects of profound neural organizations.

In exploring the general exhibition of DNNs, found a consistent outperformance all through different investigations subjects contrary to most recent non-neural

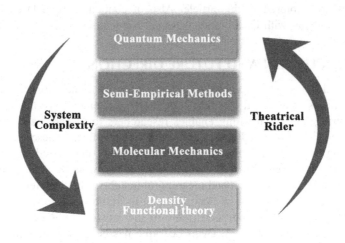

FIGURE 4.7   The basic architecture of a computational chemistry.

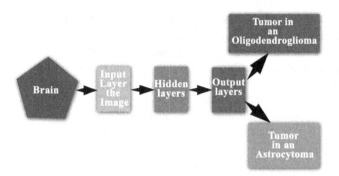

FIGURE 4.8   Deep learning to find tumor.

organizations designs, and profound neural organization essentially based thoroughly forms furthermore surpassed their particular assignments' "discriminatory constraint" assumptions. In profound dominating calculations, mixed with the multifaceted nature of GPU-raised processing for training profound neural organizations and the dramatic blast of synthetic data to show those organizations, to be a successful gadget for computational science.

4.3.13   DEEP LEARNING IN RADIOLOGY

Deep Learning makes use of interconnected networks to solve complicated problems. In a number of state-of-the-art tasks, specifically, the ones associated with images, Deep Learning algorithms have provided groundbreaking results (Figure 4.8).

TAs the clinical discipline of radiology is predicated mostly on the extraction of beneficial expertise from images, Deep Learning is a totally herbal region of utility, and studies on this discipline have developed swiftly in current years. The trendy historical past of radiology and the opportunities for the utility of Deep Learning algorithms are explored in this article. Also, it contains simple Deep Learning principles, together with Convolutional Neural Networks.

4.4   CONCLUSION AND FUTURE DIRECTIONS

In traditional fields – such as biology, medicine, and engineering – as well as in emerging fields of artificial biology, computerized chemical synthesis, and bio-manufacturing, system learning enables researchers to research and apprehend more and more complicated bodily and organic phenomena. These fields require new paradigms for the translation of more and more complicated statistics. The transition to Deep Learning and complicated modelling is an attempt to bridge the gap between complex data and meaningful insights. This chapter offers a top-level view of the application of deep learning in traditional and emerging fields and its present packages and needs. One of the most complicated and flexible branches of statistical sciences, Deep Learning has been ever-changing and has great potential.

REFERENCES

Aizenberg, I. N., Aizenberg, N. N., & Vandewalle, J. (2000). Multi-valued and universal binary neurons.

Chatfield, K., Simonyan, K., Vedaldi, A., & Zisserman, A. (2014). Return of the devil in the details: Delving deep into convolutional nets. *arXiv preprint arXiv:1405.3531*.

Dalal, N., & Triggs, B. (2005). Histograms of oriented gradients for human detection. In *2005 IEEE Computer Society Conference on Computer Vision and Pattern Recognition (CVPR'05)* (Vol. 1, pp. 886–893).

Everingham, M., Eslami, S. M. A., Van Gool, L., Williams, C. K. I., Winn, J., & Zisserman, A. (2015). The pascal visual object classes challenge: A retrospective. *Int. J. Comput. Vis.*, *111*(1), 98–136.

Deng, L., & Yu, D. (2014). Deep learning: Methods and applications found. *Foundations and Trends in Signal Processing*, *7*(3–4), 197–387.

Girshick, R. (2015). Fast R-CNN. In *Proceedings of the IEEE International Conference on Computer Vision* (pp. 1440–1448).

Girshick, R., Donahue, J., Darrell, T., & Malik, J. (2014). Rich feature hierarchies for accurate object detection and semantic segmentation. In *Proceedings of the IEEE Conference on Computer Vision and Pattern Recognition* (pp. 580–587).

He, K., Gkioxari, G., Dollár, P., & Girshick, R. (2017). Mask R-CNN. In *Proceedings of the IEEE International Conference on Computer Vision* (pp. 2961–2969).

He, K., Zhang, X., Ren, S., & Sun, J. (2016). Deep residual learning for image recognition. In *Proceedings of the IEEE Conference on Computer Vision and Pattern Recognition* (pp. 770–778).

Hinton, G. E., & Salakhutdinov, R. R. (2006). Reducing the dimensionality of data with neural networks. *Science*, *313*(5786), 504–507.

Hinton, G., Deng, L., Yu, D., Dahl, G. E., Mohamed, A-R., Jaitly, N., Senior, A., Vanhoucke, V., Nguyen, P., Sainath, T. N., & Kingsbury, B. (2012). Deep neural networks for acoustic modeling in speech recognition: The shared views of four research groups. *IEEE Signal Process. Mag.*, *29*(6), 82–97.

Koehn, P. (1994). Combining genetic algorithms and neural networks: The encoding problem.

Krizhevsky, A., Sutskever, I., & Hinton, G. E. (2017). Imagenet classification with deep convolutional neural networks. *Commun. ACM*, *60*(6), 84–90.

Kuo, C.-C. J. (2016). Understanding convolutional neural networks with a mathematical model. *J. Vis. Commun. Image Represent.*, *41*, 406–413.

LeCun, Y., Bengio, Y., & Hinton, G. (2015). Deep learning. *Nature*, *521*(7553), 436–444.

LeCun, Y., Bottou, L., Bengio, Y., & Haffner, P. (1998). Gradient-based learning applied to document recognition. *Proc. IEEE*, *86*(11), 2278–2324.

Lin, T.-Y., et al. (2014). Microsoft COCO: Common objects in context. In *European Conference on Computer Vision* (pp. 740–755).

Liu, W., Wang, Z., Liu, X., Zeng, N., Liu, Y., & Alsaadi, F. E. (2017). A survey of deep neural network architectures and their applications. *Neurocomputing*, *234*, 11–26.

Liu, W., et al. (2016). SSD: Single shot multibox detector. In *European Conference on Computer Vision* (pp. 21–37).

Liu, L., et al. (2020). Deep learning for generic object detection: A survey. *Int. J. Comput. Vis.*, *128*(2), 261–318.

O'Mahony, N., Murphy, T., Panduru, K., Riordan, D., & Walsh, J. (2017). Improving controller performance in a powder blending process using predictive control. In *2017 28th Irish Signals and Systems Conference (ISSC)* (pp. 1–6).

O'Mahony, N., et al. (2019). Deep learning vs. traditional computer vision. In *Science and Information Conference* (pp. 128–144).

Pathak, A. R., Pandey, M., & Rautaray, S. (2018). Application of deep learning for object detection. *Procedia Comput. Sci.*, *132*, 1706–1717.

Redmon, J., Divvala, S., Girshick, R., & A., Farhadi (2016). You only look once: Unified, real-time object detection. In *Proceedings of the IEEE Conference on Computer Vision and Pattern Recognition* (pp. 779–788).

Ren, S., He, K., Girshick, K., & Sun, J. (2015). Faster R-CNN: Towards real-time object detection with region proposal networks. In *Advances in Neural Information Processing Systems* (pp. 91–99).

Rosenblatt, F. (1957). *The Perceptron, a Perceiving and Recognizing Automaton Project Para*. Cornell Aeronautical Laboratory.

Rumelhart, D. E., Hinton, G. E., & Williams, R. J. (1986). Learning representations by back-propagating errors. *Nature*, *323*(6088), 533–536.

Scherer, D., Müller, A., & Behnke, S. (2010). Evaluation of pooling operations in convolutional architectures for object recognition. In *International Conference on Artificial Neural Networks* (pp. 92–101).

Szegedy, C., et al. (2015). Going deeper with convolutions. In *Proceedings of the IEEE Conference on Computer Vision and Pattern Recognition* (pp. 1–9).

Uijlings, J. R. R., Van De Sande, K. E. A., Gevers, J., & Smeulders, A. W. M. (2013). Selective search for object recognition. *Int. J. Comput. Vis.*, *104*(2), 154–171.

Voulodimos, A., Doulamis, N., Doulamis, A., & Protopapadakis, E. (2018). Deep learning for computer vision: A brief review. *Comput. Intell. Neurosci.*, *2018*.

Zhao, Z.-Q., Zheng, P., Xu, S., & Wu, X. (2019). Object detection with deep learning: A review. *IEEE Trans. Neural Networks Learn. Syst.*, *30*(11), 3212–3232.

5 Emerging Trend and Research Issues in Deep Learning with Cloud Computing

S. Karthi, D. Deepa, and M. Kalaiyarasi

CONTENTS

5.1 INTRODUCTION

Over the years, distributed computing has created a demand for different computing models, such as cloud computing. Cloud computing depends upon specialist farms. These specialist farms are generally located far from the clients. Cloud computing is developing at a fast pace. It has been proposed that many Internet of Things (IoT) gadgets would be in use in 2020.

The distance between clients and the cloud causes low correspondence idleness, influencing the nature of the association (QoS) and nature of commitment (QoE). Besides, data on client region, near to affiliation conditions and clients' adaptability lead cannot be immediately obtained.

Before distributed computing, there was the Client/Server configuration, in which all information is left on the labourer side. When a lone customer needed to express data or run a program, they interfaced with the laborer and subsequently secured fitting access, and thereafter they can do their business. Around then after, flowed enlisting came into picture, where all the PCs are coordinated together and offer their resources when required upheld above enrolling, there was emerged of disseminated processing thoughts that later realized.

5.1.1 Cloud Computing Architecture

SDCAs, the small and goliath partnership use a sufficient registration progress to supply the data in the cloud and access it from anywhere the web affiliation is used at any point. The design of disseminated processing can be a mixture of organizational strategy and event-driven design. Distributed computing architecture is divided into the two parts that follow. Front End and Back End.

The front component is used by the consumer. It requires interfaces and frameworks on the client side that are supposed to enter the distributed phases of calculation. The front fuses web staff with small and large clients and tablets (checking Chrome, Firefox, web explorer, etc.). The specialist group uses the back end. It manages all the services that disseminated registration organizations are supposed to provide.

Figures 5.1 and 5.2 shows the architecture of cloud computing.

Through the Internet, utilizing the remuneration per-use procedure, the public cloud is available to anybody for storage and accessing information. The Cloud Service Provider (CSP) oversees and works the enlistment properties in the open-air cloud (e.g., Amazon's versatile figure cloud (EC2).

Alternatively, the private cloud is otherwise referred to as an internal cloud or corporate cloud. Associations use it to assemble and deal with their own server farms within or outside the business. It is often submitted using Open source

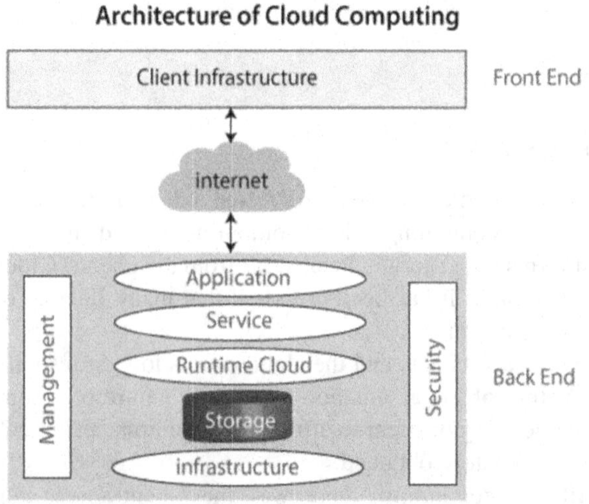

FIGURE 5.1 Cloud computing architecture.

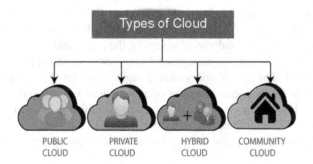

FIGURE 5.2   Cloud service types.

software, such as Open stack and Eucalyptus. A mixture of the public cloud and the private cloud is the hybrid cloud. We might say: *Hybrid Cloud = Public Cloud + Private Cloud*

A hybrid cloud is mostly safe because everyone can get the administrations that run on the public cloud, while administrations that run on a private cloud can only get customers of the association. A hybrid cloud is mostly secure.

A local area cloud permits frameworks and administrations to be open by a few associations to divide the information among the association and a chosen local area. It is possessed, oversaw, and worked by at least one association inside the local area, an outsider, or a combination of them. Distributed computing is Internet-based registering. Mists are circulated innovation stages that influence innovation developments to supply exceptionally adaptable and tough conditions. Ibrahim et al. (2019) Shared assets, programming and information are given to PCs and different gadgets on-request, much the same as the power network. Equipment, the framework program, and applications are conveyed as an administration over the web.

Under programs such as IaaS, PaaS and SaaS (Chouhan & Hasbullah, 2016) cloud handling, the cloud offers upholding that is the appropriate stage for significant learning analysis considering the way the architecture provides support for Scalability, Virtualization, Storage for huge proportions of premium data co-ordinated and unstructured and boundless resources. Flexibility is that a structure, association, or cycle is restricted to handling a proportion of work generated in a capable way. It is the biggest factor for the evaluation of datasets of beasts. The regular models spend an outsized proportion of their time arranging game plans for scale-up and scale-out. On the gear level, simple hypotheses are made. Circulated processing, by offering on-demand (adaptable) figuring tools on the fly, dispenses with this overhead for the Architect/Organization.

A part of the architecture is then reduced to the cloud model, providing flexibility for knowledge processing. It is set up to handle a lot of data and supply vital data/ yield exercises (IOPS) every second to move on data to examining tools. This provides storage for both organized and unstructured material. Thus, database adaptability, spread enrollment, and virtualization ensures that there is never a shortage of storage space (Wang & Hou, 2016).

The Cloud Computing model gives boundless resources on interest. Colossal data conditions require a gathering of laborers to help the devices that cycle tremendous volumes fast and varied setups of the organization provisioning data. In this way, they offer an appreciation of common sense to support tremendous advances in data. The use of disseminated registration for the execution of beast data eliminates the in-house handling of power duty by transferring the details to the cloud. They offer enough leeway to partnerships evaluated with little to medium reach. For example, Iaas combines taking the actual hardware and goes fully virtual, all experts, partnerships, accumulating, and device heads inside the cloud all present.

This is typically the same establishment and hardware that runs within the cloud inside the non-conveyed figuring method (Youssef et al., 2016). This may alleviate the requirement for local area data, warming, cooling, and local level hardware maintenance. This model of organization is the one that can be used to accumulate a lot of data. By rapidly sending additional handling centers, IaaS advancement ups to take care of limits, IaaS helps you to delegate or surrender mutual specialist services. These are virtualized to handle the registration and restrict the beast data evaluation requirements. Unrivaled staff, partnerships, and scarce resources are handled through cloud work schemes. Versatility allows for the rapid and similar transfer of resources, as needed.

Disseminated statistics place colossal knowledge within the reach of associations that might not somehow be set up to deal with the expense of the enormous costs or contribute to the time associated with the procurement of adequate hardware capacity to store and separate enormous enlightening assortments. Circulated statistics have a pool of mutual resources: consumers are provided. To buyers of the company from a common pool, resources such as figure, memory, affiliation, and circle (storing) are allocated. Fast adaptability has been created by enabling the provisioning and appearance of resources rapidly as interest in the cloud organization increases and decreases. Every now and then, this is an after-effect of the association.

The ability to quickly adjust helps you to get a good arrangement on figure, memory, association, and resource limits. This is also because these services are graded and transmitted when appropriate. When adventures begin, the resources are distributed to cloud organization customers, and when the effort comes to an end, these resources are transmitted back to the cloud establishment.

Count: Amazon EC2 has developed an API to dispatch enrolling events with any of the retained working systems at the required processing level. It supports the discovery of different models by Amazon Computer Images (AMIs). Google Big inquiry is a summary of how cloud expert partnerships regularly make the computing of huge data straightforward. It licenses Query organizations for outstandingly huge datasets and gives precise results quickly (Nenvani & Gupta, 2016).

5.2 DEEP LEARNING

Deep Learning is a way to deal with AI-empowered PC frameworks to improve with experience and information. It speaks to the planet as a settled pecking order of ideas, where each connection is characterized as far as easier ideas and more unique portrayals are figured regarding the normal registering and the standardless complex ones (Nenvani & Gupta, 2016).

Deep Learning has acquired ubiquity because of its headway in registering ability by the appearance of a graphics processing unit (GPU), diminished equipment cost, and an improved organization network. In addition, the multiplication of training information and, subsequently, the momentum of scientific development in AI and information handling contribute to the visible efficiency of Deep Learning.

Instead of utilizing a physically created assortment of rules to get highlights of information, Deep Learning has the ability to discover the fundamental highlights at the preparation stage. Deep Learning uses an assortment of successive layers (10 or more) with each layer giving a larger representation of the knowledge text. In testing AI fields such as image characterization, voice recognition, and penmanship record, tongue handling, self-driving cars, and bunches of everything else, it has been used.

ML is characterized as a "Field of study that gives PCs the ability to discover without being unequivocally customized. ML might be an auxiliary way of registering that enables machines to discover without unequivocal programs and subsequently the very AI. ML developed from the prevalence of examples and hypothesis of PC learning. Some vital ideas for ML and routinely utilized ML calculations for wise investigation are talked about there. The objective of directed learning is to foresee the correct vector yield for a given info vector. In cases with at least one nonstop factor inside the objective mark, relapse is perceived as relapse. It is hard to characterize the objective of solo learning. One among the most targets is to spot delicate bunches inside the information input, called the grouping, of similar examples. The consequences of ML calculation are frequently essentially improved through this handling step and are named extraction of usefulness.[m1]. See Figure 5.3.

Deep Learning is embraced over edge figuring. Since there are restricted edge hubs, we likewise plan another download technique for the enhancement of ML in edge processing. In the presentation appraisal, the exhibition of ML undertakings is tried utilizing our methodology in a high-level processing climate. The outcomes show that our strategy gives extra Deep Learning enhancement in the field of new technologies. See Figure 5.4.

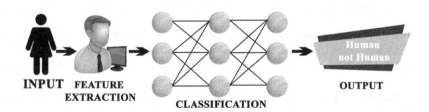

FIGURE 5.3 Machine learning.

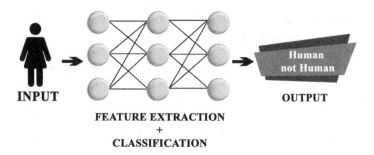

FIGURE 5.4   Deep learning.

5.2.1   Supervised and Unsupervised Learning

The two mechanisms of AI are supervised and unsupervised learning. Nevertheless, both methods are used under different environments and for different datasets. Controlled counts of learning are prepared using named data. The managed learning model involves direct analysis to verify whether the correct yield is foreseen. The yield is estimated by the supervised learning model. In controlled learning, the model is given an input record near the yield. The aim of guided learning is to coach the model altogether so that when new data is given, it can imagine the yield. Driven change allows the managers to mentor the model. For those situations where we usually interpret the data similarly as contrasting yields, coordinated learning is consistently used.

A definite outcome is the supervised learning model. Driven learning is not similar to the real edge of true AI, as we first train the model for all data during this, and then no one can predict the correct yield yet. It integrates different estimates, such as rectilinear backslide, logistic regression, vector machine support, multiclass classification, decision tree, Bayesian logic, etc. Deep Learning computations are readied using unlabeled data. The protected plans are discovered in data through an independent learning model. The aim of independent learning is to look at the disguised models and supportive pieces of data from the dark dataset. If they appeared differently in relation to supervised learning, the unsupervised learning model may provide less detailed results. It fuses multiple estimates such as the algorithm Clustering, KNN, and Apriori (Yuan & Max, 2017).

5.2.2   Deep Learning Techniques Adopted in the Emergent Cloud Environment

The use of Deep Learning in emergent conveyed processing has been found in a couple of territories from cultivating, prosperity, resource the heads, transportation, waste the board, presumption examination, object disclosure, advanced assurance, etc. Different enlisting counts are wont to evaluate the introduction inside the disseminated registering atmosphere. With a mix of Deep Learning and circulated processing, this gives the advantages of the customer experience inside the hour of organization provisioning. The overview of estimations and their inclinations are given. The various figurings include Convolutional Neural Networks, Deep Reinforcement Learning, Deep Neural Networks, Recurrent Neural Networks, and Deep Belief Networks. These count works merge with the cloud environment (Liu et al., 2017).

5.3   CONVOLUTIONAL NEURAL NETWORK

A Convolutional Neural Network (ConvNet/CNN) is a Deep Learning technique that can ingest an image of knowledge, assign significance (learnable loads and predispositions) within the image to changed viewpoints/objects, and be prepared to distinguish one from the other. When compared with other arrangement measurements, the pre-handling needed during a CNN is much lower. Although channels in crude strategies are hand-designed, CNNs have the ability to discover these channels/qualities with adequate planning. The fundamental flow of the CNN is given below. See Figure 5.5.

A CNN's engineering is similar to that of the accessibility instance of neurons within the human brain and was enlivened by the visual territory connection. The response of particular neurons increases only during a restricted region of the field of vision referred to as the Receptive Field. A bunch of such fields covers the entire visual cortex to protect it.

5.4   DEEP REINFORCEMENT LEARNING

A brisk zone, Deep Reinforcement Learning (DRL), is a mix of Reinforcement Learning and Deep Learning. It can handle a reasonable extent of complex one-of-

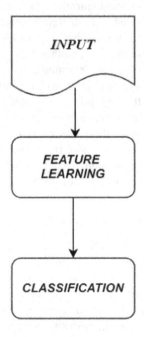

FIGURE 5.5   Convolutional neural network.

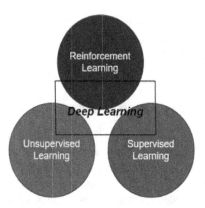

FIGURE 5.6 Deep reinforcement learning.

a-kind assignments that were previously difficult for a machine to deal with veritable issues in an application with human-like intuition.

Perhaps learning by cooperating with our environmental variables is the basic approach that involves our psychology, once we embrace the learning personality. It is the way that we interpret how a baby learns. Furthermore, that we realize that such collaborations are without a doubt a significant wellspring of information about our environmental factors and ourselves for the duration of individuals' lives, not simply newborn children. For instance, at the point when we sort out some way to energize a vehicle, we are completely aware of how the environment answers to the excuse we do and that through our exercises we likewise hope to influence what occurs in our natural factors. See Figure 5.6.

A major idea that underlies most learning speculations is to gain from the collaboration and is that the establishment of Reinforcement Learning in combination of both the space. Reinforcement Learning's approach focuses far more on objective coordinated association learning than other Machine Learning, comparatively.

The preparation substance is not determined by what activities are required, however, rather it should find for itself which activities, by testing them by "experimentation," produce the best prize, their objective. Furthermore, these activities can influence the quick compensation as well as the more extended term ones, "postponed rewards", since the current activities will decide future circumstances (how it occurs, in actuality). These two qualities, "experimentation" search and "deferred reward", are two distinctive attributes of fortification discovering that we'll cover all through this arrangement of posts.

5.5 RECURRENT NEURAL NETWORK

The Recurrent Neural Network (RNN) is a Neural Network where the yield from past improvement is treated as a pledge to the current turn of events. All data sources and yields are delivered from one another in standard neural affiliations, however in cases, for example, when it is relied upon to foresee the subsequent explanation of a sentence, the past words are required and along these lines there is a fundamental word. Henceforth, with the assistance of a hidden layer, RNN showed up, keeping an eye on this issue.

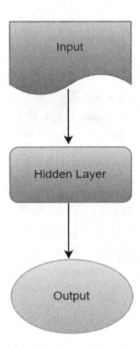

FIGURE 5.7   Recurrent neural network.

The hidden layers are the most imperative part of RNN, which recalls two or three plans for some information. RNN has a "memory" that reviews all the details of what has been resolved. See Figure 5.7.

For all data, it uses equivalent limits as it carries out an indistinguishable effort to supply the yield on all information sources or hidden layers. In contrast to other neural associations, this diminishes the multi-faceted idea of limits.

iDeep Belief Networks are a graphical portrayal, which are generative in nature, for example, it creates all potential qualities that might be produced for the current situation. It is a mixture of likelihood and measurements with AI and Neural Networks. Deep Belief Networks contains different layers with values, wherein there is a connection between the layers but not the qualities. The most important point is to help the framework arrange the information into various classifications. See Figure 5.8.

The first generation Neural Networks utilized Perceptrons, which distinguished a particular article or object by contemplating "weight" or pre-taken care of properties. The Perceptrons must be powerful at a fundamental level and not helpful for cutting-edge innovation. To disentangle these issues, the second generation of Neural Networks saw the presentation of the idea of Back spread, during which the yield is contrasted and the predefined yield, as well as the mistake esteem, are diminished to zero. Backing Vector Machines made and saw more experiments by relating to recently entered experiments. Next came co-ordinated a cyclic charts called conviction networks which helped in taking care of issues related with deduction and learning issues. This was trailed by Deep

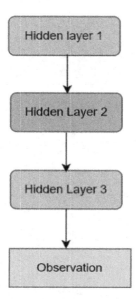

FIGURE 5.8 Deep belief network.

Belief Networks, which assisted with making fair qualities to be put away in leaf hubs.

5.6 DEEP LEARNING APPLICATIONS IN THE EMERGING CLOUD COMPUTING ENVIRONMENT

The rapid growth of the cloud and the number of mobile apps are increasing; it is also important to enhance the performance of the application, so that the combination of both cloud and Deep Learning offers better performance in the evolving world. See Figure 5.9.

5.7 CHALLENGES AND NEW PERSPECTIVE FOR FUTURE DIRECTION

In emerging circulated registration plans, the allocation of substantial learning has brought enormous piles of focal points. There are unpredictable difficulties given the eye drawn in by the expansiveness of the assessment district. The difficulties are talked about and future investigation way with new experiences is spread to advance future assessment improvement.

1. Inside the evolving circulated figuring structures, there is a multi-target enhancement problem: Some applications centered on precision achieving high torpidity also have low battery life. Others focused on inaction at the expense of exactness. In the two models, QoS is influenced differently. With low torpidity and extended battery future, contraption customers should have high accuracy. Multi-target results in these issues.

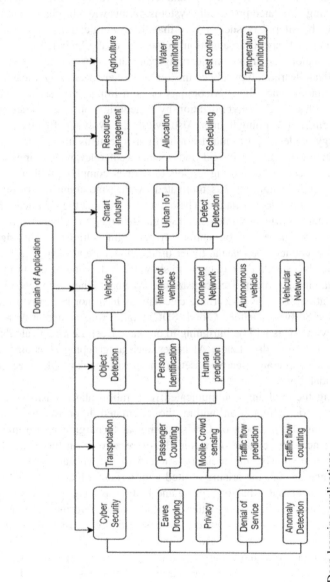

FIGURE 5.9 Deep learning applications.

2. In the region where there is no voluminous data, moving learning and performing various tasks for critical learning are regularly implemented to scale back the complexity of non-appearance of enormous data size in the zone where there is no voluminous data. In addition, we propose that examiners use moving learning and conduct various errands learning in emerging circulated processing systems. In this way, the significant learning should be altered in such a way that those applications may use critical learning that pays little attention to the size of the knowledge.

3. Combination and important learning calculation for social events: Cream-significant learning typically counts outmaneuver specific significant learning computation since the shortcomings of both important learning figures are always discarded by hybridization by benefiting from the characteristics of the significant learning figures. Also, within the evolving effective processing strategy, collecting essential learning figures remains unexploited. It is also fascinating to get the hybrid and social event relevant learning counts in emerging conveyed figuring models to unravel complex problems

4. Meta-heuristic figuring was motivated by the smoothing out of significant learning estimation by nature: The audit revealed that the significant learning computations widely implemented within the emerging circulated registering frameworks depended on manual hyper limit settings of the significant learning counts. The manual hyper limit settings are abnormal, monotonous, and needs a standard exact cycle to show up at the ideal hyper limit regards

5. Programming characterized registering stay unexploited with Deep Learning calculation: The Deep Learning calculations have not been investigated inside the Software-Defined Cloud (SDC) engineering, as demonstrated inside the review. This will be intriguing to embrace Deep Learning calculations in SDC climate to disentangle AI issues such as grouping, bunching, and the expectation for an effective asset of the board, bookinSQLg, security, and dependability.

6. Among the evolving communicated registration plans, business emphasis is eccentric competition within the circulated figuring business focus, triggering the prospect of CPUs, storing and charging a customer correspondence. The evaluation would therefore zero in on the level of virtual CPUs and, thusly, the memory allotted to the virtual CPU. As a consequence of that, the arising orbited appropriated processing configuration requires a complex business place and it stays as an open assessment issue.

7. Energy usage is limited by batteries because of the reformist running of basic learning models. The various evaluations made by such gadgets trigger high energy use, which continually prompts the fate of low batteries. When efficiently performed, huge learning tallies may prompt high energy utilization on the contraptions. When large learning calculations are carried out on the workers, the cloud worker ranch will similarly experience high energy use.

8. Lethargy problems: On resource-compliant IoT computers, when critical learning applications are performed. This will have a serious effect, particularly in the field of related vehicles and abundance the board, where little postponement could cost lives.

9. Different important learning figures with their varieties are alive: Inconvenience in choosing adequate significant learning count. Researchers/engineers conclude that it is difficult to settle on the most un-troubled Deep Learning estimate to understand the emerging disseminated processing mechanisms for handling problems. This is consistently in light of the fact that there is no depicted Deep Learning model equipment.

10. More expansive: research is needed on Deep Learning in evolving trans-mitted registration systems, as this is regularly a child research district from the evaluation neighborhood as of now. The probability of the emerging distributed figuring strategy for further enhancement of the room is expected to be fully appreciated by wider research on the allocation of substantial learning.

5.8 CONCLUSION

The problems of the standard circulated figuring guidelines animated the progress of processing structures conveyed by coming to age. The emerging dispersed fig-uring models produce large proportions of data that are beyond the ability to gauge the shallow rapid estimates. Important learning counts have begun to receive tre-mendous contemplations inside the evolving disseminated figuring scholarly works, with their ability to deal with huge degree datasets. A summary of the gathering of Deep Learning in emerging transmitted registration models is organized by and through composition. Significant training is gaining a very high premium in emerging dispersed processing and is expected to continue due to the creation of architecture and new possibilities for future evaluation. In order to include headings for tending to the perceived issues, questionable evaluation obstacles and new insight for future study orientation are addressed.

REFERENCES

Al-garadi, M. A., Hussain, M. R., Khan, N., Murtaza, G., Nweke, H. F., Ali, I., & Gani, A. (2019). Predicting cyberbullying on social media in the big data era using machine learning algorithms: Review of literature and open challenges. *IEEE Access*.

Bengio, Y. (2009). Learning deep architectures for AI. *Foundations and Trends in Machine Learning*, 2(1), 1–127.

Chiroma, H., Gital, A. Y. U., Rana, N., Shafi'i, M. A., Muhammad, A. N., Umar, A. Y., & Abubakar, A. I. (2019). Nature inspired meta-heuristic algorithms for deep learning: Recent progress and novel perspective. In *Science and Information Conference* (pp. 59–70), Cham.

Chiroma, H., Herawan, T., Fister Jr, I., Fister, I., Abdulkareem, S., Shuib, L., & Abubakar, A. (2017). Bio-inspired computation: Recent development on the modifications of the cuckoo search algorithm. *Applied Soft Computing*, 61, 149–173.

Chouhan, M., & Hasbullah, H. (2016, August). Adaptive detection technique for cachebased side channel attack using Bloom Filter for secure cloud. In *IEEE International Conference on Computer and Information Sciences* (pp. 293–297).

Dalal, N., & Triggs, B. (2005). Histograms of oriented gradients for human detection. In *IEEE Computer Society Conference on Piscataway* (pp. 886–893).

Hosein, S., & Hosein, P. (2017). Load forecasting using deep neural networks. In *IEEE Power & Energy Society Innovative Smart Grid Technologies Conference* (pp. 1–5).

Hinton, G. E., Osindero, S., & Teh, Y. W. (2006). A fast learning algorithm for deep belief nets. *Neural Computation, 18*(7), 1527–1554.

Ibrahim, T. M., Alarood, A. A., Chiroma, H., Al-garadi, M. A., Rana, N., Muhammad, A. N., & Gabralla, L. A. (2019). Recent advances in mobile touch screen security authentication methods: A systematic literature review. *Computers & Security, 85*, 1–24.

Khan, A., Chiroma, H., Imran, M., khan, A., Bangash, J.I., Asim, M., Hamza, M.F., & Aljuaid, H. (2020). Forecasting electricity consumption based on machine learning for advancing performance: The case study for the organization of petroleum exporting countries (OPEC). *Computers & Electrical Engineering – Elsevier.*

Lavi, B., FatanSerj, M., & Ullah, I. (2018). Survey on deep learning techniques for person re-identification task. *arXiv preprint arXiv:1807.05284.*

Lee H., Grosse R., Ranganath R., & Ng A. Y. (2009). Convolutional deep belief networks for scalable unsupervised learning of hierarchical representations. In *Proceedings of the 26th Annual International Conference on Machine Learning* (pp. 609–616), New York, NY, United States.

LeCun, Y., Bengio, Y., & Hinton, G. (2015, May). Deep learning. *Nature, 521*(7553), 436–444.

Leng, Q., Mang Y., & Qi T. (2019). A survey of open-world person re-identification. *IEEE Transactions on Circuits and Systems for Video Technology.*

Liu, J., Osadchy, M. Foster, M., Ashton, L., Solomone, C.J., & Gibson, S.J. (2017). Deep convolutional neural networks for Raman spectrum recognition: A unified solution. *Analyst, 142*, 4067–4074.

Luo, P., Tian, Y., Wang, X., & Tang, X. (2014). Switchable deep network for pedestrian detection. In *Proceedings of International Conference on Computer Vision and Pattern Recognition* (pp. 899–906), New York, NY, United States.

Nenvani, G., & Gupta, H. (2016, March). A survey on attack detection on cloud using supervised learning techniques. In *IEEE Symposium on Colossal Data Analysis and Networking* (pp. 1–5), Indore, India.

Ogabe, T., Ichikawa, H., Sakamoto, K., et al. (2016). Optimization of decentralized renewable energy syes-Asia (ISGT -Asia) (pp. 1014–1018).

Ouyang, W., Chu, X., & Wang, X. (2014). Multi-source deep learning for human pose estimation. In *Proceedings of International Conference on Computer Vision and Pattern Recognition* (pp. 2337–2344).

Wang, K., & Hou, Y. (2016, October). Detection method of SQL injection attack in cloud computing environment. In *IEEE Advanced Information Management, Communicates, Electronic and Automation Control Conference* (pp. 487–493).

Wang, P., Li, W., & Ogunbona, P., et al. (2018). RGB-D-based human motion recognition with deep learning: A survey. *Computer Vision and Image Understanding, 171*, 1–22.

Wang, K., Wang, H., Liu, M., Xing, X., & Han, T. (2018). Survey on person re-identification based on deep learning. *CAAI Transactions on Intelligence Technology, 3*(4), 219–227.

Wu, D., Zheng, S.-J., Zhang, X.-P., Yuan, C.-A., Cheng, F., Zhao, Y., & Huang, D.-S.. (2019). Deep learning-based methods for person reidentification: A comprehensive review. *Neurocomputing, 337*, 354–371.

Youssef, B. C., Nada, M., Elmehdi, B., & Boubker, R. (2016, November). Intrusion detection in cloud computing based attacks patterns and risk assessment. In *International Conference on Systems of Collaboration* (pp. 1–4).

Yuan, Y., & Max, Q.-H. (2017). Deep learning for polyp recognition in wireless capsule endoscopy images. *Medical Physics, 44*, 1379–1389.

Zeng, X., Ouyang, W., & Wang, X. (2013). Multi-stage contextual deep learning for pedestrian detection. In *Proceedings of International Conference on Computer Vision and Pattern Recognition* (pp. 121–128). IEEE.

6 An Investigation of Deep Learning

S. Karthi, P. Kasthurirengan, M. Kalaiyarasi, D. Deepa, and M. Sangeetha

CONTENTS

6.1 INTRODUCTION

Deep Learning is a core area of Machine Learning, for algorithms to find solutions to complex, unrecognized, and unpredictable problems using available datasets, or to make reflections. Deep Learning (DL) uses layers of figuring to procedure information, see human talk, and ostensibly catch objects. Information is passed through a couple of layers. The fundamental level is known as the data level, and the extra is known as a yield layer. All the layers collectively are called hidden layers. Each layer is regularly a fundamental, uniform game plan of rules containing one kind of establishment. DL strategies that replicate human neural networks have become well known due to the expansion of high-in general execution registering office. DL accomplishes higher force and adaptability because of its capacity to handle an enormous number of highlights when it manages unstructured information.

DOI: 10.1201/9781003038450-6

DL learns traditional instructions and permits the information through many layers; each layer can mine capabilities step by step and pass them to the subsequent layer. Initial layers extract low-degree capabilities, and succeeding layers combine capabilities to shape an entire representation.

The records of DL may be traced back to 1943 when Walter Pitts and Warren McCulloch made advancement reliant on NNs of the human brain (PalmAd, 1984). They used a mix of estimations and calculating they called "edge reasoning" to reflect the human thought (Michalík, 2009; Tariq et al., 2020). Since that time, DL has developed reliably, with the best huge breaks in its new development. Henry J. Kelley is credited for building up the fundamentals of a persistent Back Propagation (BP) Model in 1960. During the year 1962, a less complicated model primarily based totally simplest at the chain rule become evolved with the aid of using Stuart Dreyfus. While the idea of BP did exist in the early 1960s, it was clumsy and inefficient, and did not grow to be beneficial until1985.

6.1.1 ARTIFICIAL NEURAL NETWORK

To delineate the construction of the artificial neural network (ANN), a functional and valuable appearance is necessary to be obtained from the human mind. The mind incorporates roughly 10^{11} processing units, "neurons", working in equal and changing data by means of their connectors, "neurotransmitters"; those neurons summarize all the data getting into them, and if the outcome is superior to the given ability alluded to as movement capacity, they send a heartbeat through the axon to the ensuing stage (Smith, 2011).

In an equivalent manner, counterfeit neural local area incorporates simple processing devices, "fake neurons," and an individual unit is joined to the elective contraptions through bulk connections; at that point, those devices compute the weighted amount of the moving toward inputs and find the yield the utilization of crushing trademark. Figure 6.1 demonstrates the synthetic neuron.

In light of the basic structure and capacity of the NN, three essential components of a neuron prototype can be recognized:

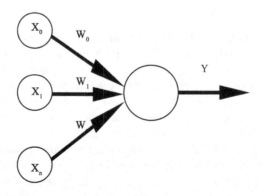

FIGURE 6.1   Basic function of artificial neuron.

1. Synapses, or partner joins, have a mass or power where the data signal x_i related with neuron k is expanded by synaptic weight wi.
2. A snake for adding the weighted information sources.
3. An incitation ability to make the yield of a neuron. It is moreover implied as a devastating limit, in that it squashes (confines) the ampleness extent of the caution sign to a restricted worth.

Exactly, the productivity on the neuron y_1 can be designated as

$$y = \varphi \sum w_i x_i$$

where

- $x_1, x_2, x_3, \ldots, x_n$ are the input's signals.
- $w_1, w_2, w_3, \ldots, w_n$ are the respective weights of neuron.
- φ is the activation function.

6.1.2 CONVOLUTION NEURAL NETWORKS

The first Convolutional Neural Networks (CNNs) were used by Kunihiko Fukushima. Fukushima planned NNs with an increased number of grouping together of convolutional layers (Fukushima, 1987). In 1979, he advanced an ANN, alluded to as Neocognitron (Fukushima, 1979), which utilized a various leveled, multi-layered plan. See Figure 6.2.

The Neocognitron become animated with the guide of utilizing the form proposed with Hubel and Wiesel in 1959. From the information, noticed kinds of blocks in the graphic essential cortex known as unpretentious and multifaceted block, also initiate a falling adaptation of such blocks to be utilized in Patten identification of the design tasks (Fukushima, 2007; Li et al., 2012). The network looked like current adaptations, anyway had been talented with a support approach of repeating actuation in different layers, which got strength over the long haul. Moreover, Fukushima's design permitted pivotal capacities to be changed physically with the guide of utilizing developing the "weight" of the connection.

6.1.3 NEOCOGNITRON

The Neocognitron is a progressive, multi-layered ANN created by Kunihiko Fukushima in 1979 (Haykin, 2009). It has been utilized for transcribing Japanese individual standing and distinctive patten identification task, and filled in as the establishment for CNN. See Figure 6.3.

The Neocognitron is a homegrown augmentation of those falling models. The Neocognitron is made out of more than one style of blocks, the greatest fundamentals of which are known as S-blocks and C-blocks. The local highlights are removed by S-block, and those capacities twisting, which incorporate nearby moves, are endured by C-blocks. Nearby capacities in the info are incorporated

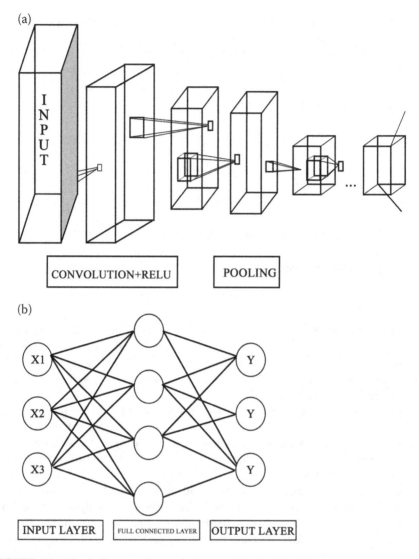

FIGURE 6.2   Block diagram of convolution neural networks.

consistently and named with inside the advanced layers (Fukushima, 1987). The idea of neighborhood work reconciliation is resolved in a few unique models, which incorporate the CNN model, the SIFT strategy, and the HoG technique.

6.1.4 BACK PROPAGATION

Back Propagation (BP), the usage of a goof in getting ready DL models, advanced significantly in 1970. This happened while Seppo Linnainmaa wrote his Master's thesis, alongside a FORTRAN program for again multiplication. Appallingly, the idea become now not, now completed to NN till 1985. This happened while

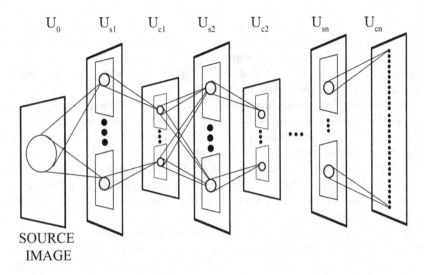

U_0 U_{s1} U_{c1} U_{s2} U_{c2} U_{sn} U_{cn}

SOURCE
IMAGE

FIGURE 6.3 Basic structure of neocognitron.

Rumelhart, Williams, and Hinton showed again expansion in a Neural Network to give "interesting" movement depictions. Judiciously, this disclosure passed on to smooth the request inside mental cerebrum study of whether human data relies upon meaningful reasoning (computationalism) or allocated depictions (connectionism). In 1989, Yann LeCun furnished the fundamental sensible presentation of BP at Bell Labs. He blended CNN in with again multiplication onto break down "Transcribed" digits.

This contraption was finally used to examine the amounts design examination. Likewise, while the subsequent AI winter (1985–1990s) was presented, this additionally influenced assessments for NN and DL. Distinctive exorbitantly significant people had distorted the "speedy" furthest reaches of AI, breaking expectations and bothering examiners. The inconvenience became so remarkable, the articulation AI showed up at pseudoscience status. Fortunately, a few people endured to materials on AI and DL, and two or three full-size impels have been made. In 1995, Dana Cortes and Vladimir Vapnik built up the guide vector machine (a system for arranging and spotting comparative data).

LSTM (extended short period of time period memory) for RNN become progressed in 1997, through strategy for strategies for Sepp Hochreiter and Juergen Schmidhuber. The ensuing huge developing advance for DL happened in 1999, while PC frameworks began transforming into faster at preparing records and graphical processing units (GPU) were created. Quicker handling, with GPUs preparing pictures, sped up by the method of methods for 1,000 occasions longer than a year's range. During this time, NN headed out to rival uphold vector machines. While an NN could be moderate contrasted with a help vector machine, NNs offered higher outcomes the utilization of similar records.

NNs, moreover, have the addition of driving forward to improve as the preparing information is added. During 1999–2000, one problem was created called

"Vanishing Gradient". It turned into the identification of "features" shaped in initial layers had been now no longer being found out by the upper layers, due to the fact no gaining knowledge of sign reached those layers. This turned into now no longer an essential hassle for all NNs, simply those with gradient-primarily based totally gaining knowledge of methods.

Two answers have been used to remedy this hassle were stage-by-stage pre-modeling and the improvement of lengthy primary memory. Around 2011–2012, Google Brain began a phenomenal undertaking perceived as The Cat Experiment. In free-vigorous task researched the issues of "unaided learning (USL)." DL uses "regulated learning (SL)," which suggests the CNN is capable of the use of requested data (expect pics from ImageNet). Using USL, a convolutional CNN is given unlabeled data, at that point referenced to searching for out routine models.

The Cat Experiment utilized a NN unfold more than 1000 PCs. 10 million "unlabeled" pictures are taken from haphazardly in online streaming websites, demonstrated to the framework, after which the tutoring programming pattern changed into permitted to execute. That time quit of the schooling, one neuron in the greatest stage changed into found to answer unequivocally to the pictures of the cat. Andrew Ng, the assignment's organizer stated, "We also found a neuron that answered firmly to human appearances." USL stays an enormous reason in territory of DL.

This experiment stabilizes around 70% of its trailblazers in preparing unnamed pictures. Nonetheless, it analyzed substantially 16% lesser than the contraptions utilized for schooling, also surprisingly more terrible with devices that had been convoluted or change. As of now, the preparation of Big Data and the development of AI are each settled on DL. DL remains in need of development and creative ideas.

6.1.5   BACKPROPAGATION NEURAL NETWORK ARCHITECTURE

The Back Propagation Neural Network engineering, includes the areas of the building plan, execution estimation, work guess capacity, and learning. The utility of the BP methodology in setting up a suitable load in a disseminated versatile network has been indicated over and again (Vogl et al., 1988). Unfortunately, in various applications, the number of accentuations required before blend can be colossal. Changes to the BP calculation portrayed by Rumelhart et al. (1985). See Figure 6.4.

- Instead of reviving the network load after every model is acquainted with the framework, the framework is invigorated simply subsequent to the whole assortment of guides to be erudite have been acquainted with the association, at which time the logarithmic measures of all the mass modified are functional:
- As opposed to having η, the "learning rate" steady, it is contrasted vigorously so the count use a nearby ideal η, as directed by the local improvement topography;

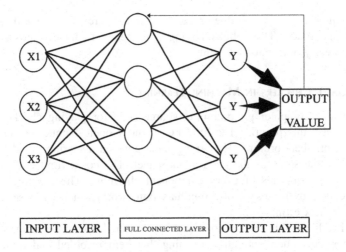

INPUT LAYER   FULL CONNECTED LAYER   OUTPUT LAYER

FIGURE 6.4 Backpropagation neural network architecture.

- The energy factor α is locate to nil while, as meant by a breakdown of a phase to diminish the whole slip-up, the information trademark in previous advances will undoubtedly be misdirecting than significant. Solely after the framework makes a supportive pace, i.e., individual to diminishes the hard and fast bungle, does α again anticipate a non-zero worth. Contemplating the assurance of weight in neural nets as an issue in conventional nonlinear upgrade speculation, the thinking for figurings searching for simply those heaps that produce the inside and out the world least mix-up is investigated and excused.

6.2 HISTORY OF MACHINE LEARNING

Machine Learning (ML) is a basic issue of present-day business endeavor and examination. It utilizes algorithm and brain cell community models to help PC frameworks in improving their presentation. ML calculations regularly build a numerical rendition of the use of test information – furthermore recognized as "prepared data" – to settle on decisions without being uncommonly modified to settle on the one's decisions.

ML is, to some degree, fundamentally dependent on a variant of psyche mobileular association. The rendition got made in 1949 with the guide of utilizing Donald Hebb, his hypotheses on neuron invigoration and verbal trade among brain cells. Hebb expressed, "When one cell consistently helps with terminating another, the axon of the main cell creates synaptic handles (or augments them on the off chance that they as of now exist) in contact with the soma of the subsequent cell." Modifying Hebb's norms to an engineered brain structure and manufactured brain cells, his adaptation might be characterized as a way of changing the connections among manufactured neurons and the alterations to character neurons. The seeking among cell/hubs fortifies if the two cells/hubs are enacted on the indistinguishable time and debilitates on the off chance that they're actuated independently. "Weight" is utilized to clarify those connections, and hubs/cells having a tendency to be each fine or each horrendous are characterized as having powerful fine loads. Those hubs having a tendency to have opposite weight increment hearty horrendous weight.

6.2.1 GAME CHECKERS IN MACHINE LEARNING

IBM created a computer programming source code for playing checkers in the 1950s. Because this program has little PC memory open, Samuel began what is known as alpha-beta pruning. His arrangement covered a scoring brand name the utilization of the spots of the parts at the panel. The count limit endeavored to measure the possibilities of every component charming. The program picks its resulting course by the usage of a minimax framework, which likewise advanced into the minimax estimation.

Samuel additionally arranged some parts permitting his item to emerge as better. In what Samuel called reiteration learning, his item recorded and reviewed all positions it had viably clear and mixed this in with the assessments of the prize limit. Arthur Samuel recently showed up with the articulation "ML" in 1952. Algorithm for Supervised Learning of Binary Classifiers (Perceptron):

The Binary Classifiers changed into in the first place conscious as a framework, presently not, at this point a program. The product program, from the outset intended for the IBM 704, changed into mounted in a specially developed framework, known as the Mark 1 Binary Classifiers, which have worked for image recognition. This made the product program and the calculations adaptable and to be had for various machines.

Depicted on the grounds that the initial hit a cell system, the Mark I Binary Classifiers progressed a couple of issues with harmed assumptions. Although the Binary Classifiers seemed promising, they could not comprehend numerous types of obvious examples (counting faces), causing disappointment and slowing down neural local area considers. It very well may be various years sooner than the dissatisfactions of purchasers and venture bunch blurred. Neural people group/ Machine Learning contemplates battled till resurgence all through the 1990s.

6.2.2 ALGORITHM FOR NEAREST NEIGHBORS (k-NN)

During 1967, the k-NN set of rules became considered, which become the beginning of essential example acknowledgment. This arrangement of rules became used for planning courses and became probably the soonest calculation used in finding a technique to the venturing salesclerk's difficulty of finding the greatest green course. Utilizing it, a salesclerk enters a particular city and on numerous occasions has this framework go to the nearest towns till all had been visited. Marcello Pelillo has been given FICO rating for developing the "closest neighbor rule."

6.2.3 FORWARDING INFORMATION BETWEEN LAYERS

During the 1960s, the creation of multiple layers created a brand in a different direction in new AI invention. AI became useful for offering and for the use of a new stage to the Binary Classifiers, it provided considerably extra working strength than the Binary Classifier uses of a single block. From another variation of NN had been created after the Binary Classifier opened the door to "layers" in system, and the form of NN keeps expanding. The creation of more than one layer provides information to NN and BP.

6.2.4 ARTIFICIAL NEURAL NETWORK (ANN)

An ANN has hidden layers that may be used to answer additional mind-boggling commitments than the older Binary Classifier could. ANNs are the main gadget used for ML. Neural Networks use enter and yield layers and, typically, are comprised of a hidden layer (or layers) intended to change over go into insights that might be utilized the through method of methods for yield layer. The hiddenb layers are top-notch for finding styles excessively confounded for a human developer to distinguish, which implies that a human could not find the example after which to instruct the device to get it. See Figure 6.5.

ANNs are naturally invigorated PC bundles intended to recreate the way that the human brain receives data. ANNs collect their skill through distinguishing the styles and connections in records and learn (or are prepared) through experience, presently no longer from programming. An ANN is formed from stacks of unmarried units, fabricated neurons, or getting ready parts (PE), associated with coefficients (loads), which address the neural shape and are composed in layers.

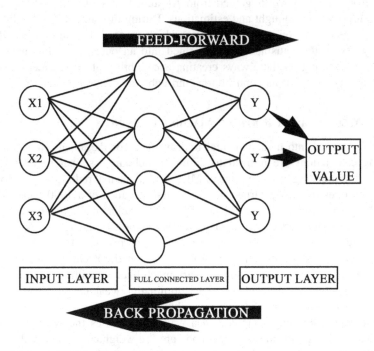

FIGURE 6.5 Artificial neural network.

The strength of AI calculations comes from interfacing neurons locally. Every PE has weighted sources of info, switch trademark, and one yield. The lead of an AI local area is chosen through the switch abilities of its neurons, through the considering rule, and through the actual design. The loads are the flexible boundaries and, in that sense, an AI local area is a defined framework. The gauged amount of the data sources comprises the initiation of the neuron. The enactment sign is outperformed through a change trademark to give an unmarried yield of the neuron. Move trademark acquaints non-linearity with the local area. During preparation, the again spread calculations are utilized to streamline the mistake, forecasts are limited, and the local area arrives at the ideal level of exactness (Agatonovic-Kustrin & Beresford, 2000).

6.2.5   Machine Learning versus Artificial Intelligence

During the 1970s and 1980s, AI researchers had focused on the utilization of sensible information fundamentally based absolutely methodologies in inclination to calculations. Furthermore, AI people group considers transformed into abandoned through PC innovative skill and AI scientists. This created a split between AI and Machine Learning. Up to that point, ML had been used as a coaching programming for AI.

The AI attempt, which covered a monster measure of researchers and specialists, changed into upgraded straightforwardly into an alternate subject and fought for practically 10 years. The endeavor direct moved from mentoring for AI toward fixing sensible issues in articulations of giving commitments. Its mindfulness moved from the frameworks gained from AI studies to strategies and techniques utilized in possibility thought and estimations. During this time, the ML adventure kept up its perception on AI networks after which succeeded withinside the 1990s. The majority of this satisfaction transformed into an outcome of Internet development, capitalizing on the always creating accessibility of virtual measurements and the ability to extent its contributions through way of the Internet.

6.2.6   Algorithm for Boosting Machine Learning

"Boosting" turned into a fundamental improvement for the advancement of ML. Boosting calculations are utilized to diminish inclination sooner or later of administered dominating and comprise of ML calculations that redesign weak novices into hearty ones. Boosting turned out to be first provided in a 1990 paper named "The Strength of Weak Learnability," by Robert Schapire. Schapire states, "A bunch of weak beginners can make an unmarried vigorous student." Weak amateurs are depicted as classifiers, which are best scarcely corresponded with the veritable arrangement (by the by higher than arbitrary speculating). On the other hand, a powerful student is easily arranged and very much lined up with the real grouping.

Most algorithms are created for dreary dominating weak categorization, which at that point transfer to an absolute last robust categorizer. In the wake of being added, they are normally weighted in a way that assesses the weak beginner's precision. At that point, the realities loads are "re-weighted." Input realities this is misclassified benefits a superior weight, simultaneously as realities arranged effectively sheds pounds. These environmental factors a grant predetermination weak beginner's to acknowledgment extra radically on going before weak amateur's that have been misclassified.

The straightforward differentiation among the various sorts of algorithms is "the technique" used for weight calculation realities focuses. AdaBoost is an acclaimed Machine Learning set of rules and generally huge, being the essential arrangement of rules ready to running with weak beginner. An enormous assortment algorithm works of art in the AnyBoost structure.

6.2.7 FACIAL MODEL IDENTIFICATION

During the year 2006, the Facial Model Identification Challenge – a National Institute of Standards and Technology program – evaluated the upheld Facial Model Identification calculation of the time. 3-D face checks, iris pictures, and significant standard face pictures were attempted. Their disclosures suggested that the new counts were on numerous occasions more accurate than the customized face affirmation figurings from 2002, and, on various occasions, more exact than those from 1995. A portion of the estimations were set up to beat human individuals in recognizing appearances and will exceptionally perceive vague twins. In 2012, Google's X Lab developed an ML computation, which freely examines and finds accounts containing cats. In 2014, Facebook made one calculation prepared for seeing or affirming individuals in photographs with vague accuracy as individuals.

6.2.8 HOW MACHINE LEARNING HAPPENED TODAY?

ML was portrayed by Stanford University as "the investigation of getting PCs to act without being explicitly adjusted." Machine Learning is, as of now, responsible for presumably the fundamental degrees of progress in development, for instance, the new business of self-driving vehicles. Artificial Intelligence has actuated another assortment of thoughts and advances, including managed and solo learning, new counts for robots, the Internet of Things, assessment devices, chatbots, and that is just a glimpse of something larger. Recorded underneath are seven fundamental ways the universe of business is as of now using ML:

- Analyzing Sales Data: Streamlining the information
- Real-Time Mobile Personalization: Promoting the experience
- Fraud Detection: Detecting design changes
- Product Recommendations: Customer personalization
- Learning Management Systems: Decision-production programs
- Dynamic Pricing: Flexible valuing dependent on a need or interest
- Natural Language Processing: Speaking with people

Artificial Intelligence models have gotten adaptable in perseveringly acknowledging, which makes them logically accurate the more they work. ML computations got together with new preparing developments advance flexibility and improve efficiency. Paired with a business examination, Machine Learning can resolve an

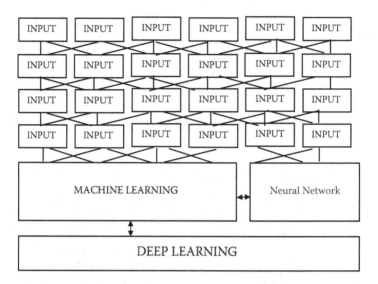

FIGURE 6.6   Block diagram of deep learning.

arrangement of progressive complexities. Presently, ML models can be used to make predictions regarding disease diagnosis to the rising and fall of stocks.

6.3   DEEP LEARNING

Deep NNs are, for the most part, deciphered as far as the general estimate theorem (Cybenko, 1989) or probabilistic inference (Deng & Yu, 2014). The exemplary widespread guess hypothesis concerns the limit of feed-forward NNs with one hidden layer of limited size to estimated consistent functions (Cybenko, 1989). In 1989, the fundamental proof was conveyed by George Cybenko for sigmoid incitation capacities (Cybenko, 1989), and was summarized to deal with forwarding multi-layer plans in 1991 by Hornik (1991). Recent work, furthermore, showed that broad assessment in like manner holds for non-restricted inception limits like the changed direct measure (Sonoda & Murata, 2017).

The boundless assessment speculation for significant NNs concerns the constraint of associations with restricted width yet the significance is allowed to create. Lu et al. (2017; Murphy, 2012; Hinton 2012; Vinyals et al., 2014; Krizhevsky, 2012; Krizhevsky et al., 2012; Oh & Jung, 2004) showed that if the width of a significant NN with ReLU commencement is cautiously greater than the data estimation, by then the association can be construed by any Lebesgue integrable limit; If the width is more unassuming or acceptable the data. See Figure 6.6.

6.4   CONCLUSION

Deep Artificial NNs (for example, intermittent ones) have acquired a few challenges in example acknowledgment and AI. This old overview minimalistically sums up pertinent work, a great deal of it from the former thousand years. Shallow and Deep Learning models are conspicuous through methods for the force of their FICO rating undertaking ways, which can be chains of likely learnable, causal connections among developments and impacts. I review profound directed learning (furthermore restating the records of BP), solo learning, support learning and developmental calculation, and angled search for fast bundles encoding profound and gigantic Network.

REFERENCES

Agatonovic-Kustrin, S., & Beresford, R. (2000). Basic concepts of artificial neural network (ANN) modeling and its application in pharmaceutical research. *Journal of Pharmaceutical and Biomedical Analysis, 22*(5), 717–727.

Cybenko. (1989). Approximations by superpositions of sigmoidal functions. *Mathematics of Control, Signals, and Systems, 2*(4), 303–314. doi:10.1007/bf02551274

Deng, L., & Yu, D. (2014). Deep learning: Methods and applications. *Foundations and Trends in Signal Processing, 7*(3–4), 1–199. doi:10.1561/2000000039

Fukushima, K. (1979, October). 位置ずれに影響されないパターン認識機構の神経回路のモデル --- ネオコグニトロン --- [Neural network model for a mechanism of pattern recognition unaffected by shift in position — Neocognitron —]. *Trans. IECE (in Japanese),* J62-A(10), 658–665.

Fukushima, K. (1987). A hierarchical neural network model for selective attention. In R. Eckmiller & C. Von der Malsburg (Eds.), *Neural computers* (pp. 81–90). Germany: Springer-Verlag.

Fukushima, K. (2007). Neocognitron. *Scholarpedia, 2*(1), 1717. doi:10.4249/scholarpedia.1717

Haykin, S. (2009). *Neural networks and learning machines* (3rd ed., p. 938). Hamilton, ON, Canada: Pearson Education Inc.

Hinton, G. E., Srivastava, N., Krizhevsky, A., Sutskever, I., & Salakhutdinov, R. R. (2012). Improving neural networks by preventing co-adaptation of feature detectors. *arXiv:1207.0580 [math.LG].*

Hornik, K. (1991). Approximation capabilities of multilayer feedforward networks. *Neural Networks, 4*(2), 251–257. doi:10.1016/0893-6080(91)90009-t

Krizhevsky, A., Sutskever, I., & Hinton, G. (2012). ImageNet classification with deep convolutional neural networks. *NIPS 2012: Neural Information Processing Systems,* Lake Tahoe, Nevada.

Li, J., Cheng, J.-H., Shi, J.-Y., & Huang, F. (2012). Brief introduction of back propagation (BP) neural network algorithm and its improvement. In *Advances in computer science and information engineering* (pp. 553–558). Berlin, Heidelberg: Springer.

Lu, Z., Pu, H., Wang, F., Hu, Z., & Wang, L. (2017). The expressive power of neural networks: A view from the width. *Neural Information Processing Systems,* 6231–6239.

Michalík, J. (2009). *Applied neural networks for digital signal processing with DSC TMS320 F28335.* New York, NY, USA: Technical Univerzity of Ostrava.

Murphy, K. P. (2012, August 24). *Machine learning: A probabilistic perspective.* Cambridge, MA: MIT Press.

Oh, K.-S., & Jung, K. (2004). GPU implementation of neural networks. *Pattern Recognition, 37*(6), 1311–1314.

PalmAd, G. (1984, October 1–4). *Aertsen, brain theory – Proceedings of the tirst trieste meeting on brain theory.* ISBN-978-3-642-70911-1.

Rumelhart, D. E., Hinton, G. E., & Williams, R. J. (1985). Learning internal representations by error propagation (No. ICS-8506). San Diego La Jolla Institute for Cognitive Science, California University.

Smith, S. W. (2011). Human hearing. In *The Scientist and Engineer's Guide to Digital Signal Processing*.

Sonoda, S., & Murata, N. (2017). Neural network with unbounded activation functions is universal approximator. *Applied and Computational Harmonic Analysis*, *43*(2), 233–268.

Tariq, M. I., Memon, N. A., Ahmed, S., Tayyaba, S., Mushtaq, M. T., Mian, N. A., & Ashraf, M. W. (2020). Survey – A review of deep learning security and privacy defensive techniques.

Vinyals, O., Toshev, A., Bengio, S., & Erhan, D. (2014). Show and tell: A neural image caption generator. *arXiv:1411.4555* [*cs.CV*].

Vogl, T. P., Mangis, J. K., Rigler, A. K., Zink, W. T., & Alkon, D. L. (1988). Accelerating the convergence of the back-propagation method. *Biological Cybernetics*, *59*(4–5), 257–263.

7 A Study and Comparative Analysis of Various Use Cases of NLP Using Sequential Transfer Learning Techniques

R. Mangayarkarasio, C. Vanmathi, Rachit Jain, and Priyansh Agarwal

CONTENTS

DOI: 10.1201/9781003038450-7

7.1 INTRODUCTION

Natural language is used by humans for everyday communication. Every human language is evolving by addressing its own inherent challenges and ambiguities. Natural language processing is a domain that presents various text manipulation techniques for human languages. Because the development of human language is a continuous process, it is inevitable for linguistics and technicians to present new kinds of techniques to complement the process (Alqaryouti et al., 2019). The usage of the internet by more users has produced a huge amount of data.

Analyzing data and inferring insight has become a challenging task for data analysts (Habibi et al., 2017; Seon et al., 2001; Wu et al., 2001). In the literature, many researchers contributed various Machine learning and Deep Learning techniques to solve the various use cases of the natural language problem. Among the various use cases of NLP, the Sentiment Analysis (SA) and Named Entity Recognition (NER) have become the most crucial application in the trend (Wu et al., 2017; Zhang et al., 2018; Zhu et al., 2018). The objective of sentiment analysis is to get to know a user's or audience's opinion about a target object by analyzing a huge amount of textual information from various sources (Cruz et al., 2016). The insight gaied by sentiment analysis from the vast amount of data is used in many applications (Mehdiyev et al., 2017; Tripathy et al., 2015). The objective of the NER system is to extract entities from raw data and determine their corresponding category. This information is useful in a variety of NLP tasks such as Information Extraction Systems, Question-Answer Systems, Machine Translation Systems, Automatic Summarizing Systems (Fang & Zhan, 2015), and Semantic Annotation (Choi & Lee, 2012). Segmenting the various entities and establishing relationships among them is a tedious task (Siering et al., 2018). Though a reasonable amount of effort is put into identifying the various entities, there is still a requirement for domain-based entity segmentation (Kang & Zhou, 2017; Riccardo et al., 2017; Zhou et al., 2014).

The researchers analyzed the sentiments from the raw text on three levels, the document level (Yessenalina et al., 2010), sentence level (Appel et al., 2016), and aspect level (Ruder et al., 2016). Sentiment analysis or classification at the degree of the document is to predict an overall sentiment to an opinion document. The input to the system is textual information conveyed in the document, and the expected output is the measure of the insight in terms of positive or negative. In the literature, many researchers address the sentiment analysis at the level of the document and have found that document representation plays an important role. Most of the authors use the BOW model (Johnson & Zhang, 2015; Zhai & Zhang, 2016) to represent textual data, whereas the technique (Tang et al., 2015) learns sentence representation using Neural Network architectures, such as a CNN or an LSTM, from word embedding. Then, it uses GRU to adaptively encode the semantics of sentences and their inherent relations. It uses word embedding (Zhou et al., 2016) to represent text and design an LSTM architecture for cross-lingual sentiment classification. Some of the authors (Wang and Xia, 2017; Wang et al., 2016) address the sentiment analysis at the degree of sentence level, which is finding the sentiment expressed in the given sentence. In Socher et al. (2013), the authors presented the

Recursive Neural Tensor Network (RNTN) to understand the relations between elements. In Vo & Zhang (2015), the authors applied an LSTM for Twitter sentiment classification. The third category of the technique (Dahou et al., 2016; Dong et al., 2014; Li and Lam, 2017) addresses the sentiment at the degree of the target aspect. Such a category of technique considers the sentiment and target aspect from the text. This section presents various approaches to solve SA and NER problems using Machine learning and Deep Learning models.

The major problem of Machine Learning is that the models are highly dependent on large amounts of high-quality data. Unfortunately, data is rarely available and highly expensive to access. Transfer learning facilitates the less requirement of high-quality data. Transfer learning is a focused model that trains a task for which labeled training data is enormous and handles similar tasks with fewer data. These pre-trained models are often faster than our traditional prototypes. So, the focused models that are trained will henceforth contribute to the SOTA for a variety of tasks. This chapter addresses the trends of the SA and NER problems and demonstrates the various solutions derived through machine learning and transfer learning models, along with an insight into how transfer learning influences the tasks. Section 2 address the existing works, Section 3 discuss the background of the research work, Section 4 discusses the impact of transfer learning for the NLP problems of SA and NER. Section 5 is the conclusion.

7.2 LITERATURE REVIEW

This section discusses the bench-mark works contributed by various authors using transfer learning to address a variety of NLP applications. The authors in Pan & Fellow (2009) have classified the recent trends of transfer learning into three broad categories, Inductive transfer, Transductive transfer, and Unsupervised transfer. The authors inadequately elaborated on the unsupervised learning concepts. When prior assumptions, such as the source and target domains, are not related to each other and do not hold well, that may lead to negative transfer happening. The authors claimed that their techniques may find application in social networking analysis and video classification. The methods used in Alqaryouti et al. (2019) can extract the applicable streaming approaches from the sentiment classification. Henceforth, it is established that the approaches used, including lexicons and other rules, can be handled in sentiment analysis. The work may be improvised using proper classification and categorization.

The technique by Chen (2019) is mainly used to evaluate across datasets and to keep the approach in a stream of networks, such as Bidirectional-Long Short-Term Memory (BiLSTM) networks. The biggest challenge in clinical NLP is that incorrect assertions can cause incorrect diagnoses of patients, which can further lead to major issues. To prevent this, the authors proposed a model for assertion detection.

The authors, Fei & Zheng (2010), proposed a domain adaptive transfer learning method and established transfer tests among a different corpus to increase the effectiveness. The authors used only a few strategies to acquire transfer knowledge and have less linguistic information for transfer learning. The method by

Francis et al. (2019) pre-trains a model for a certain NER task and then fine-tunes the concerned learned model for another NER task, where later models have few labeled training data. To accomplish this task, the technique used word embedding, attention mechanisms, and the BERT language model.

In Lee et al. (2016), they proposed ANN and NER with the different transfer learning approaches and the parameters have been labeled according to the dataset available, as per the human observations. Rodriguez et al. (2018) showed experimental results with different corpus pairs (source/destination) of networks. They analyzed three existing methods that are applied to the setting of transfer learning with novel entities in the target domain. However, these methods have not been compared to one another previously in the literature.

Daval-frerot & Moreau (2018 and Sun et al. (2018) introduced the technique for detecting valence tasks using transfer learning BLSTM technology. To avoid overfitting, layers of the pre-trained model were frozen. The system primarily focused on single transfer, instead of on multiple transfers, to increase the amount of data used in the process. Slowly, the system may migrate from binary classes to seven classes. Zafarian et al. (2015) identified the various corpora pairs and showed the best unlabeled bilingual corpora systems. Further, they extracted the features by transferring information from another resource-rich language.

Liang et al. (2018) conducted experiments on different corpora data items, which resulted in a transferring network system for the betterment of sentence classification. Henceforth, the implemented approach is well suited for transferring into the Chinese corpus from the English corpus. Results showed that the network transferring system is quite well-functioning and gives the best performance with fewer annotations of models. The less usage of annotations is the drawback of this method. Medhat et al. (2014) present a complete picture of SA techniques and its related fields with brief details. The rich observations made by the authors are about the detailed study on the various sentiment analysis approaches and their working behavior with the corpora systems.

Liu & Shi (2019) summarizes the appropriate system modules and its neighboring entities for sentiment analysis. It focuses on the corpora pairs and fits the outcome of the English language with the particular assigned entity. Dai & Le (2015) presented the corpora entities to the lower level of text analysis and then formed numerous networks. It has been observed that the unremarkable networks provide a detailed study on sentiment analysis by a clear language model. LSTM is noted for the easy processing of identified words and also replicates the chosen corpora data items.

Peters et al. (2018) demonstrated the contextualized word representations in the sentimental approach by gathering the data items of a specific language. It mainly deals with the six challenging NLP problems that exhibit the process of deep internal models of an improvised system. Henceforth, the system guarantees the functioning of the ELMO model that resembles the process of sentiment analysis.

Howard & Ruder (2018) showcased the system performance and results of the considered datasets. It mainly results in the reduced error rate of the considered module by 18% to 24% on the selected data items. It also functions on the other data items, to a large extent. Here, the robust earning approach on the majority of datasets is found that the efficiency is good and that corpora entities are also working with the same kind of data models. Bahdanau et al. (2014) reveal the usage of models in sentiment analysis and bring the results of each part of the search system. They further notice the prediction of words and other related items becomes complex because the language uses new items that are not regularly available in the dictionary. Addressing unknown and rare words is required for the model to be more useful.

Feilmayr (2011) combined the benefits of data mining and information extraction methods. The aim was to provide a new, high-quality information extraction method and, at the same time, to improve the performance of the underlying extraction system. Thus, they shorten the life cycle of information extraction engineering. Precise feature extraction is obtained through data mining methods, which, in turn, reduces the feature space to the most significant information for mining new knowledge. Zhan & Jiang (2019) summarize the tasks and related concepts of event extraction, analyzing, comparing, and generalizing the relevant descriptions in different fields. The article reveals that entity identification, relationship identification, and syntax analysis results lead to cascading errors.

In Dawar et al. (2019), the authors explored the topic of modeling with the use of various approaches and convenient tools, such as python libraries. The datasets considered are initially pre-processed, and then the corresponding entities are extracted as the named entities. Furthermore, the characterized corpora words are aligned towards the sentiment analyzer.

This section briefly discusses the application of transfer learning for various applications. Previously, those tasks were implemented using machine learning techniques, and, more recently, using Deep Learning architectures. One of the major problems of machine learning is that the models are highly dependent on large amounts of high-quality data. Unfortunately, these datasets are rarely available, and highly expensive to access. With the help of transfer learning, there is less need for high-quality data. Transfer learning is a pre-trained model that has already been trained on a task for which labeled training data is enormous, and which can handle similar tasks with less data. These pre-trained models are often faster than our traditional machine learning models and can contribute to the SOTA in a variety of tasks. This chapter demonstrates the transfer learning models in designing the predictive modeling tasks, such as SA and NER.

7.3   EMPIRICAL STUDY

In the past few years, transfer learning gained application in object detection and some of the tasks of NLP. The main highlight of transfer learning is that it reduces the burden of designing a classifier from scratch. In other words, transfer learning helps researchers reuse or customize the model, which has been designed for some other task, to their required task. By doing so, it reduces a lot of time that needs to be spent on development and training. In contrast to traditional learning, a new model needs to be generated for every new task as shown in Figure 7.1. Transfer learning is mainly divided into transductive and inductive learning. Then, it is further narrowed into various subcategories to tackle some variants, such as domain

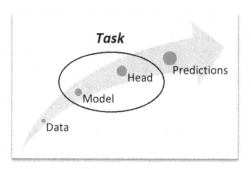

FIGURE 7.1   Task learning using the traditional model.

adaption, cross-lingual learning, multi-task learning (MTL), and sequential transfer learning (STL). This section aims to analyze the impact of sequential transfer learning in solving the benchmark models, Sentiment Analysis and NER. As the name implies, STL transfers the insight sequentially to accomplish the tasks. Sequential transfer learning is the method that has led to the biggest improvements so far. The general practice is to pre-train representations on a large, unlabeled text corpus using your method of choice and then to adapt these representations to a supervised target task using labeled data, as seen below in Figure 7.2.

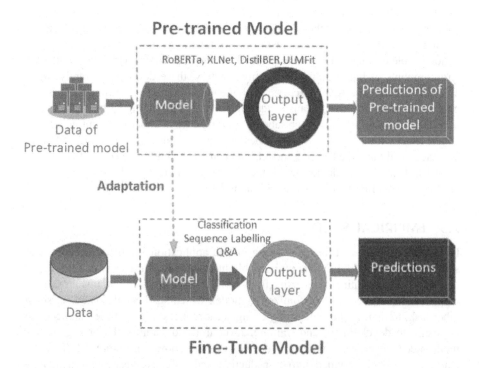

FIGURE 7.2   Sequential transfer learning process.

The target task is accomplished using sequential transfer learning (STL) in two phases, pretraining and adaptation. The pretraining phase of STL is costlier than the adaption phase since more training is required, because it is a one-time execution on the source model. The pretraining is accomplished in three approaches, distant supervision, traditional supervision, and no supervision. In distant supervision, data is obtained from heuristics and domain expertise. Traditional supervision requires manually annotated training samples, and no supervision requires a large number of unannotated samples. In adaption, a phase is accomplished through two methods, feature extraction and fine-tuning. Feature extraction uses the representations of a pre-trained model and feeds it to another model, while fine-tuning involves the training of the pre-trained model on the target task. The feature extraction and fine-tuning phase can be represented as mentioned in Eq. 7.1 and Eq. 7.2.

$$\eta_t^l = 0, \forall\, l \in [1, L_s]\ \forall\, t \tag{7.1}$$

$$\eta_t^l > 0, \exists\, l \in [1, L_s]\ \exists\, t \tag{7.2}$$

Especially, sequential transfer learning is useful in the following cases, where the process was represented with the incoming and outgoing tasks in the same period. The required models will be predicted at later stages of the sentiment process.

7.4 SEQUENTIAL TRANSFER LEARNING MODEL FOR SENTIMENT ANALYSIS

This section discusses the various STL models in the process of designing SA applications. The timeline of various transfer learning models is displayed (Figure 7.3).

FIGURE 7.3 Timeline of various models.

The benchmarking models ULMFIT, RoBERTa, XLNET, and DistilBERT have been exploited well to influence the task performance. The highlights of STL models are as follows.

7.4.1   ULMFIT

Jeremy Howard and Sebastian Ruder ( ) designed a language model, called ULMFIT. It is a method with more than just embeddings and contextualized embeddings. The ULMFIT language model can be fine-tuned for various NLP tasks. Andrew and Quoc (2015) introduce an LM that requires a huge amount of data to attain good performance. In contrast, ULMFIT completes the task with a small corpus. ULMFIT is more effective than the other TL model, known as ELMO, which makes use of the Language Model in the Fine-Tuning Process. Readers interested in the ULMFIT model should refer to (Howard & Ruder, 2018) for more information.

7.4.2   RoBERTa

RoBERTa was established at Facebook. RoBERTa uses 1000% more textual data than the BERT model, as well as computing power. To improve the training, RoBERTa eliminates the Next Sentence Prediction (known as NSP) task from BERT pre-training and introduces another masking, such as dynamic masking, so that the masked token during each epoch keeps on changing. RoBERTa uses 160GB of text for pre-training, which includes the 16GB of Books Corpus English that Wikipedia used in BERT (Reddy et al., 2020).

7.4.3   XLNet

XLNet is an improvised version of the BERT model, which has more computation power than BERT and increased efficiency in terms of accuracy. XLNet works on a different architecture than the BERT model, it uses permutation language modeling. In this model, tokens are taken and predicted randomly; because of this variability in the architecture, the throughput of learning bidirectional relationships increases, thereby enhancing the dependencies to deal with and relations between the words. It is observed that XLNet is an enhanced version of the BERT Model, as BERT is an autoencoder (AE) language. It aims to reconstruct the sentence using Mask. This model perceives the context in both forward and backward directions. However, this has several disadvantages. First, it will cause fine-tune disparency, as the mask variable is not present during the fine-tuning process in our dataset. The second is that the masked tokens assume that they are independent of other tokens in the sentences. To overcome those things, XLnet uses permutation Language Modeling, which provides better results than that. For training, this whole model uses about 130GB of data.

7.4.4   DistilBERT

DistilBERT is a small, fast in execution, economically cheap, and light Transformer model trained by distilling the BERT base. This model has 40% fewer parameters than BERT-base-uncased, and runs 60% faster, while preserving over 95% of Bert's performances. DistilBERT is the most crowdsourced platform and has numerous applications. It mainly fetches the records of data in the supervised and unsupervised data models. It also batches the different layers in dynamic masking.

7.4.5 METHODOLOGY

The steps used to implement SA using STL are given as pseudo-code and the same is depicted in the architecture diagram. The architecture diagram precisely narrates the steps that are used for the proposed SA framework. The pseudo code for STL model with pre training is given below. Apply the word-embeddings with LSTM or machine learning functions to summarize the results.

Pseudo Code 1:

Input: IMBD and Yelp reviews dataset
Output: Classifier using STL performance results
Step 1: Accept each sample from the corpus
Step 2: Tokenize the sample
Step 3: Sentence splitting
Step 4: Pre-train transfer learning model
Step 5: Adopt the pre-trained model on the target task
Step 6: Evaluate the model
Step 7: Summarize the performance using quantitative metrics

7.4.6 RESULTS AND DISCUSSION

To test the robustness as well as the efficiency of the framework using STL, the authors believe that by using the classification metrics based on the confusion matrix found in Reddy et al. (2020), it is feasible to establish the quantitative metrics, such as True positive (TP) and False positive (FP), True Negative (TN), False Negative (FN), Precision, sensitivity, specificity, accuracy, F1-score, and the execution time. Although these metrics may able to assess the designed framework and produce the performance results quantitatively, it is always necessary to check the reliability of AI frameworks. In this chapter, the authors use accuracy and considered the amount of time to complete the task to evaluate the SA and NER.

$$Accuracy = \frac{(TP + TN)}{(TP + TN + FP + FN)} \qquad (7.3)$$

The analyzed framework is tested in the python environment that is the Kaggle kernel and the default aspects of the GPU 2020. The python library version is also the default version of the Kaggle kernel library. No changes to the version were made. The learning rate of the designed Neural Network models is at the rate of $2*10^-3$ with up to 3 epochs. No additional constraints were used in NN and other ML approaches. The reason for choosing accuracy and time is for the purpose of using a transfer learning model to simulate any NLP task in less time. The speed or the amount of time to complete the required task is given more priority than other aspects of the frameworks. SA using STL is tested with two bench-marking datasets such as IMDB and Yelp-reviews using binary classification model.

7.4.6.1 Experiment Set 1 on IMDB

The IMDB dataset consists of 50k movie reviews, where the two category variables positive and negative share an equal amount of space. The results obtained through various STL models, as well as the machine and Deep Learning models, are summarized in Table 7.1. The main upside of preferring the transfer learning model is to design the task model with less time (Reddy et al., 2020). Table 7.1 shows the execution time of various models, while considering 50% of the sample and 100% of the samples. By observing the data from Table 7.1, among the various STL models, the DistilBERT works faster than other STL models. Whereas in the case of the statistical machine learning model category that is among Naive Bayes, SVM (support vector Machine), Logistic Regression, Naïve Bayes works faster than all others (Luque et al., 2018). Table 7.1 conveys that the LSTM architecture takes more time than its counterpart models, both in the case of considering 50% and 100% of the samples. Additionally, the execution time of various models used for simulating SA is displayed as a graph in Figure 7.4. In Table 7.2, the various models' performance in terms of accuracy is shown. The interpretation is that the LSTM architecture produces better accuracy for SA than the STL and Statistical Machine learning models. Among the STL models, ULMFIT produces better accuracy than RoBERTa, XLnet, and DistilBERT; the graphical interpretation can be found in Figure 7.5.

TABLE 7.1

Comparative Results of Various STL Models versus Traditional Machine/Deep Learning Model in Terms of Time

(In Minutes)	RoBERTa	XLNet	DistilBERT	ULMFit	LSTM	Naïve Bayes	SVM	Logistic Regression
100% of Dataset Time	30:40	27:41	15:13	93:04	274:32	03:12	30:13	13:41
50% of Dataset Time	13:41	14:37	07:33	49:31	111:23	01:46	09:14	06:48

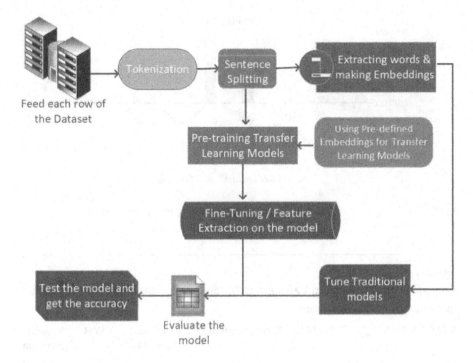

FIGURE 7.4   Sentiment analysis using STL.

TABLE 7.2
Comparative Results from Various STL Model versus Traditional Machine/Deep Learning Model in Terms of a Metric

	RoBERTa	XLNet	DistilBERT	ULMFit	LSTM	Naïve Bayes	SVM	Logistic Regression
100% of Dataset Accuracy	0.904	0.870	0.871	0.922	0.945	0.866	0.903	0.842
50% of Dataset Accuracy	0.892	0.863	0.871	0.914	0.949	0.856	0.898	0.834

7.4.6.2   Experiment Set 2 YELP Review Dataset

Yelp has been one of the most famous review sites for local businesses. Yelp mainly focuses on long-form reviews rather than short one-liners, to provide a deeper insight. The dataset contains 10,000 values, which are labeled from 1 to 5, according to user reviews. The execution setup remains the same as with SA using the IMDB dataset. The performance results of SA on Yelp reviews using STL and Machine learning models are summarized in Table 7.3. The results reveal that the

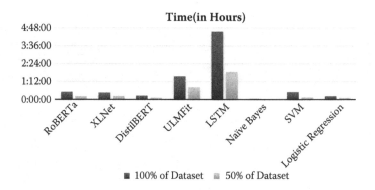

FIGURE 7.5   Time comparison between transfer learning model and traditional model.

TABLE 7.3

Comparative Results from Various STL Model versus Machine Learning Model in Terms of Execution Time

(In Minutes)	RoBERTa	XLNet	DistilBERT	Multinomial Naïve Bayes	SVM	Logistic Regression
100% of Dataset Time	01:59	09:20	01:56	04:12	04:13	4:00
50% of Dataset Time	01:05	04:55	01:00	02:16	02:16	01:56

DistilBERT model outperforms the other models, which have been considered for the analysis. The DistilBERT completes the classification task by 1m and 56s, which is 3s faster than the RoBERTa model. Among the machine learning models, the model using Logistic Regression completes the task faster than the others. The graphical representation for this is shown in Figure 7.6.

Table 7.4 summarizes the results of STL models with other machine learning models in terms of the benchmark metric called accuracy. In this context, the machine learning model outperforms the STL model. Nevertheless, our objective is to complete the task with a minimal amount of time. From this perspective, STL models perform well. The visual representation of SA on Yelp in terms of accuracy can be seen in Figure 7.7.

7.5 SEQUENTIAL TRANSFER LEARNING MODEL FOR NER

Named Entity Recognition (NER) is a necessary part of NLP tasks, such as IR and IE. NER is used to find the entity-type of words in a given dataset. This section demonstrates the NER framework using STL's two benchmarking models, ELMO and BERT. NER uses sequential transfer learning models, like BERT and Elmo, along with the basic models that use feature vectors and

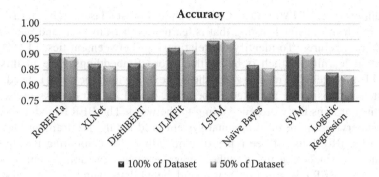

FIGURE 7.6 Comparison between transfer learning model and traditional model in terms of accuracy.

TABLE 7.4

Comparative Results from Various STL Model versus Machine Learning Model in Terms of Accuracy

	RoBERTa	XLNet	DistilBERT	Multinomial Naïve Bayes	SVM	Logistic Regression
100% of Dataset Accuracy	0.59	0.59	0.53	0.7694	0.59	0.785
50% of Dataset Accuracy	0.58	0.59	0.528	0.754	0.59	0.774

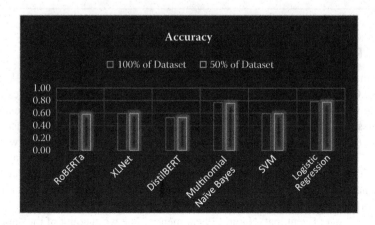

FIGURE 7.7 Execution time comparison between transfer learning model and machine learning models.

embeddings (e.g. LSTM, LSTM-CRF, Random Forest Classifier, etc.). ELMO is an approach in transfer learning that is learned both from front and back using LSTM architectures. To identify the relationships between entities, ELMO uses contextual learning, which is better than word Embeddings. Due to this, the ELMO learns the word meaning and the context in which the word is used. That is, instead of assigning the fixed embedding, the ELMO first looks in the sentence then assigns the word embedding to the word. The BERT model was first the unsupervised, pure bi-directional system used for the pre-training model of NLP tasks. BERT uses three types of embeddings for computing its input representations, including token embeddings, segment embeddings, and position embeddings. BERT is pre-trained with unlabeled data, which can be fine-tuned on labeled data to get the desired results. The advantage of BERT is that it was built on trained contextual representations and it is purely bi-directional. Whereas ELMO and ULMFIT are uni-directional and partially bi-directional. The pseudo code 2 and 3 gives the steps for detecting the NE using ELMO and BERT. 3.

Pseudo Code: 2 NER detection using ELMO model

Input: GNB data-set

Output: STL model performance

Step 1: Read the data from GNB corpus

Step 2: Pre-process the data as required for ELMO model

Step 3: Include *residual LSTM* network with an *ELMo embedding* layer.

Step 4: Fit the model

Step 5: Summarize the Model performance.

7.5.1 Results and Discussion

To facilitate the demonstration, the authors used the NER dataset, which is taken from the GMB corpus (Dai & Le, 2015). This dataset contains four variables called sentence number, words in a sentence (distributed in a row for specific sentence), parts of speech tagging, and the tags (target attribute). Target attribute uses BIO notation ('O' is used for non-entity tokens). The execution environment which is used for NER is the same as SA using STL. The GNB dataset is annotated and tagged. The analyzed framework uses the content in GNB to train as well as to predict named entities. These considered entities are as follows geo: Geographical Entity, org: Organization, per: Person, gpe: Geopolitical Entity, tim: Time indicator, art: Artifact, eve: Event, nat: Natural Phenomenon. The word count of the data set is 1,354,149. In terms of efficiency in detecting the named entities for the considered data set, Random Forest works faster than Pipeline and Deep Learning architectures. The STL model BERT works faster than ELMO.

Pseudo Code: 3 NER Detection using BERT Model

Input: GNB data-set

Output: STL model performance

Step 1: Read the data from GNB corpus

Step 2: Pre-process the data as required by the BERT model

Step 3: Define data loaders

Step 4: Use Bertfortoken class for tokenization

Step 5: Fine tune the BERT model by adjusting the parameters

Step 6: Fit the model

Step 7: Evaluate the model

The same scenario is observed while considering 100% of the data set. The results of the execution time in terms of seconds in using various ML and STL models while considering the 50% of the dataset (in Figure 7.8) and execution time is taken for 100% of the dataset (in Figure 7.9). The numerical data for the same can be found in Table 7.5. For the same dataset, STL produces a better number for accuracy as compared to other models. The graphical representation of the same can be found in Figure 7.10.

From the above discussion, the authors believe that the transfer learning model acts as a ready-to-use kit to design any classifier model. In most cases, the results obtained through the Sequential learning model are better than the earlier Machine and Deep Learning models, which may take days to complete the prediction. Discussing its merits, the researchers need to look into its downsides, too.

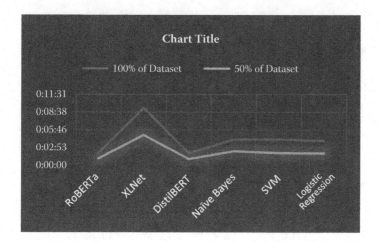

FIGURE 7.8 Execution time comparison between transfer learning model and machine learning models.

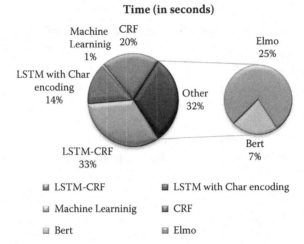

FIGURE 7.9   Time comparison between transfer learning model and traditional model in 50% dataset.

TABLE 7.5

Comparative Results from Various STL Model versus Machine Learning Model in Terms of Execution Time

(In Seconds)	Bert	Elmo	LSTM-CRF	CRF	LSTM with Char Encoding	Random Forests	ML Pipeline
100% dataset Time	1114	3966	5309	3303	2315	112	3142
50% dataset Time	617	1383	2624	1719	1074	54	1069

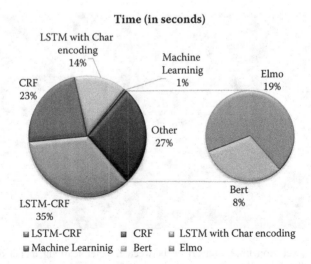

FIGURE 7.10   Time comparison between transfer learning model and traditional model in 100% datasets.

Although the transfer learning model completes the task with less amount of information, researchers should ensure the quality of data provided, the task domain, and the data that has been used to pre-train the model.

7.6 CONCLUSION

In this chapter, sequential transfer learning models as well as traditional models to simulate SA and NER tasks of NLP are examined. The proposed framework is tested using the Stanford IMDB dataset and Yelp reviews for predicting the sentiments. The Named Entity Recognition framework is tested on the GNB annotated corpus. From the results and discussion derived through STL models for SA and NER, these pre-trained models can give high accuracy with less data and, in most cases, traditional models fail to do so. While addressing the STL model for NLP, tasks need to focus on the problem of negative transfer learning. In all cases, researchers should carefully ensure the transferability between the source and the target domain.

7.7 CONFLICT OF INTEREST

None Declared

ACKNOWLEDGMENT

None Declared

REFERENCES

Alqaryouti, O., Siyam, N., Abdel, A., & Shaalan, K. (2019). Applied computing and informatics aspect-based sentiment analysis using smart government review data. *Applications of Computer Informatics, 1,* 225–235.

Andrew, M., & Quoc, V. L. (2015). Semisupervised sequence learning. *Advances in Neural Information Processing Systems* (NIPS'15).

Appel, O., Chiclana, F., Carter, J., & Fujita, H. (2016). A hybrid approach to the sentiment analysis problem at the sentence level. *Knowledge-Based Systems, 108,* 110–124.

Bahdana, D., Cho, K., & Bengio, Y. (2015). Neural machine translation by jointly learning to align and translate. In *2015 ICLR Conference.*

Chen, L. (2019). Assertion detection in clinical natural language processing: A knowledge-poor machine learning approach. In *2019 IEEE 2nd International Conference on Computer Science and Information Technology* (pp. 37–40).

Choi, Y., & Lee, H. (2012). Data properties and the performance of sentiment classification for electronic commerce applications. *Information Systems Frontiers, 19*(5), 993–1012.

Cruz, N. P., Taboada, M., & Mitkov, R. (2016). A machine-learning approach to negation and speculation detection for sentiment analysis. *Journal of the Association for Information Science and Technology, 67*(9), 2118–2136.

Dahou, A., Xiong, S., Zhou, J., Haddoud, M. H., & Duan, P. (2016). Word embeddings and convolutional neural network for Arabic sentiment classification. In *Proceedings of the International Conference on Computational Linguistics (COLING 2016)* (pp. 2418–2427).

Dai, A. M., & Le, Q. V. (2015). Semi-supervised sequence learning. In *Advances in Neural Information Processing Systems* (pp. 3079–3087).

Daval-frerot, G., & Moreau, A. (2018). Epita at SemEval-2018 task 1: Sentiment analysis using transfer learning approach. In *Proceedings of 12th International Workshop on Semantic Evaluation* (pp. 151–155).

Dawar, K., Samuel, A. J., & Alvarado, R. (2019, April). Comparing topic modeling and named entity recognition techniques for the semantic indexing of a landscape architecture textbook. In *2019 Systems and Information Engineering Design Symposium (SIEDS)* (pp. 1–6).

Dong, L., Wei, F., Tan, C., Tang, D., Zhou, M., & Xu, K. (2014). Adaptive recursive neural network for target-dependent Twitter sentiment classification. In *Proceedings of the Annual Meeting of the Association for Computational Linguistics (ACL 2014)* (pp. 242–249).

Fang, X., & Zhan, J. (2015). Sentiment analysis using product review data. *Journal of Big Data, 2*(1), 1–14.

Fei, G., & Zheng, D. (2010). Research on domain-adaptive transfer learning method and its applications. In *2010 International Conference on Asian Language Processing* (pp. 162–165).

Feilmayr, C. (2011, August). Text mining-supported information extraction: An extended methodology for developing information extraction systems. In *22nd International Workshop on Database and Expert Systems Applications* (pp. 217–221).

Francis, S., Van Landeghem, J., & Moens, M. (2019). Transfer learning for named entity recognition in financial and biomedical socuments. In *2019 IEEE 2nd International Conference on Computer and Information Technology* (pp. 1–17).

Habibi, M., Weber L., Neves M., Wiegandt, D. L., & Leser, U. (2017). Deep learning with word embeddings improves biomedical named entity recognition. *Bioinformatics, 14*, 37–48.

Howard, J., & Ruder, S. (2018). Universal language model fine-tuning for text classification. *arXiv preprint arXiv:1801.06146.*

Howard, J. & Ruder, S. (2018). Universal language model fine-tuning for text classification. *arXiv preprint arXiv:1801.06146.*

Johnson, R., & Zhang, T. (2015). Effective use of word order for text categorization with convolutional neural networks. In *Proceedings of the Conference of the North American Chapter of the Association for Computational Linguistics: Human Language Technologies (NAACL-HLT 2015)* (pp. 1412–1058).

Kang, Y., & Zhou, L. (2017). Rub, Rule-based methods for extracting product features from online consumer reviews. *Journal of Information Management, 54*(2), 166–176.

Lee, J. Y., Dernoncourt F., & Szolovits, P. (2016). Transfer learning for named-entity recognition with neural networks. *2019 IEEE 2nd International Conference on Computer and Information Technology* (pp. 4470–4473).

Li, X., & Lam, W. (2017). Deep multi-task learning for aspect term extraction with memory interaction. In *Proceedings of the Conference on Empirical Methods on Natural Language Processing (EMNLP 2017)* (pp. 502–510).

Liang, L., Zheng J., & Fu, J. (2018). Sentence classification with transfer network. In *2018 International Conference on Information Systems and Computer Aided Education* (pp. 379–382).

Liu, R., & Shi, Y. (2019). A survey of sentiment analysis based on transfer learning. *IEEE Access, 7*, 85401–85412.

Luque, A., Gómez-Bellido J., Carrasco A., & Barbancho, J. (2018). Optimal representation of anuran call spectrum in environmental monitoring systems using wireless sensor networks. *Sensors, 18*(6), 1803.

Medhat, W., Hassan, A., & Korashy, H. (2014). Sentiment analysis algorithms and applications: A survey. *AIN Shams Engineering Journal, 5*(4), 1093–1113.

Mehdiyev, N., Lahann, J., Emrich, A., Enke, D., Fettke, P., & Loos, P. (2017). Time series classification using deep learning for process planning: A case from the process industry. *Procedia Computer Science, 114*, 242–249.

Pan, S. J., & Fellow, Q. Y. (2009). A survey on transfer learning. *IEEE Transactions on Knowledge and Data Engineering, 10*, 1345–1359.

Peters, M. E., Neumann, M., Iyyer, M., Gardner, M., Clark, C., Lee, K., & Zettlemoyer, L. (2018). Deep contextualized word representations. *arXiv preprint arXiv:1802.0536*.

Reddy, G. T., Bhattacharya, S., Ramakrishnan, S. S., Chowdhary, C. L., Hakak, S., Kaluri, R., & Reddy, M. P. K. (2020). An ensemble based machine learning model for diabetic retinopathy classification. In *2020 International Conference on Emerging Trends in Information Technology and Engineering (ic-ETITE)* (pp. 1–6).

Reddy, G. T., Reddy, M. P., Lakshmanna, K., Kaluri, R., Rajput, D. S., Srivastava, S., & Baker, T. (2020). Analysis of dimensionality reduction techniques on big data. *IEEE Access, 8*, 54776–54788.

Riccardo, M., Wang, F., Wang, S., Jiang, X., & Joel, T. (2017). Deep learning for healthcare: Review, opportunities and challenges. *Briefings in Bioinformatics, 19*(6), 1236–1246.

Rodriguez, J. D., Caldwell, A., & Liu, A. (2018). Transfer learning for entity recognition of novel classes. In *Proceedings of the 27th International Conference on Computational Linguistics* (pp. 1974–1985).

Ruder, S., Ghaffari, P., & Breslin, J. G. (2016). A hierarchical model of reviews for aspect-based sentiment analysis. In *Proceedings of the Conference on Empirical Methods on Natural Language Processing* (pp. 1974–1985).

Siering, M., Deokar, A. V., & Janze, J. (2018). Disentangling consumer recommendations: Explaining and predicting airline recommendations based on online reviews. *Decision Support Systems, 107*, 52–63.

Socher, R., Perelygin, A., Wu, J. Y., Chuang, J., Manning, C. D., Ng, A. Y., & Potts, C. (2013). Recursive deep models for semantic compositionality over a sentiment treebank. In *Proceedings of the Conference on Empirical Methods on Natural Language Processing (EMNLP 2013)* (pp. 1631–1642).

Sun, P., Yang, X., Zhao, X., & Wang, Z. (2018). An overview of named entity recognition. In *2018 International Conference on Asian Language Processing* (pp. 273–278).

Tang, D., Qin, B., & Liu, T. (2015). Document modelling with gated recurrent neural network for sentiment classification. In *Proceedings of the Conference on Empirical Methods in Natural Language Processing (EMNLP 2015)* (pp. 1422–1432).

Tripathy, A., Agrawal, A., & Rath, S. K. (2015). Classification of sentimental reviews using machine learning techniques. *Procedia Computer Science, 57*, 821–829.

Vo, D. T., & Zhang, Y. (2015). Target-dependent twitter sentiment classification with rich automatic features. In *Proceedings of the Internal Joint Conference on Artificial Intelligence (IJCAI 2015)* (pp. 1347–1353).

Wang, L., & Xia, R. (2017). Sentiment Lexicon construction with representation learning based on hierarchical sentiment supervision. In *Proceedings of the Conference on Empirical Methods on Natural Language Processing (EMNLP 2017)* (pp. 502–510).

Wang, J., Yu, L., Lai, R. K., & Zhang, X. (2016). Dimensional sentiment analysis using a regional CNN-LSTM model. In *Proceedings of the Annual Meeting of the Association for Computational Linguistics (ACL 2016)* (pp. 225–230).

Wu, Y., Jiang M., Lei J., & Xu, H. (2001, March). Named entity recognition in Chinese clinical text using deep neural network. *Studies in Health Technology and Informatics, 216*, 624–631.

Wu, Y., Jiang, M., Xu, J., Zhi, D., & Xu, H. (2017). Clinical named entity recognition using deep learning models. *AMIA Annual Symposium Proceedings, American Medical Informatics Association, 2017*, 1812–1819.

Yessenalina, A., Yue, Y., & Cardie, C. (2010). Multi-level structured models for document-level sentiment classification. In *Proceedings of the 2010 Conference on Empirical Methods in Natural Language Processing* (pp. 1046–1056).

Zafarian, A., Rokni A., & Ghiasifard, S. (2015). Semi-supervised learning for named entity recognition using weakly labeled training data. In *International Symposium on Artificial Intelligence and Signal Processing* (pp. 129–135).

Zhai, S., & Zhang, Z. M. (2016). Semisupervised autoencoder for sentiment analysis. In *Proceedings of AAAI Conference on Artificial Intelligence (AAAI 2016)* (pp. 1394–1400).

Zhan, L., & Jiang X. (2019, March). Survey on event extraction tBLSTMechnology in information extraction research area. In *2019 IEEE Conference 3rd Information Technology, Networking, Electronic and Automation Control Conference (ITNEC)* (pp. 2121–2126).

Zhang, K., Wang, S., & Liu, B. (2018). Deep learning for sentiment analysis: A survey. *Wiley Interdisciplinary Reviews: Data Mining and Knowledge Discovery, 8*(4), e1253.

Zhou, X., Menche, J., Laszl, A., & Sharma, A. (2014, June). Human symptoms–disease network. *Nature Communications, 5*, 4212.

Zhou, X., Wan X., & Xiao, J. (2016). Attention-based LSTM network for cross-lingual sentiment classification. In *Proceedings of the Conference on Empirical Methods in Natural Language Processing (EMNLP 2016)* (pp. 247–256).

Zhu, Q., Li, X., Conesa A., & Pereira, C. (2018). GRAM-CNN: A deep learning approach with local context for named entity recognition in biomedical text. *Bioinformatics, 34*, 1547–1554. https://www.kaggle.com/yelp-dataset/yelp-dataset/kernels

8 Deep Learning for Medical Dataset Classification Based on Convolutional Neural Networks

S. Nathiya and R. Sujatha

CONTENTS

8.1 INTRODUCTION

Deep Learning (DL) has become one of the top areas of scientific study, which also include healthcare within its applications and plays a vital role in patient monitoring. In the medical imaging field, one of the most significant problems is

DOI: 10.1201/9781003038450-8

medical dataset classification based on images, which, in turn, need to be classified into different categories (Begoli et al., 2019). Doctors use these classification images to diagnose patients as well as for research purposes. In the past, doctors used their specialized skills to identify or diagnose diseases, where the process is complex, tedious, and time-consuming (Kim et al., 2017). Now, researchers have published many articles regarding the medical classification field and related applications, which has resulted in positive outcomes. Medical image classification typically has two steps: the first is extracting useful features from the medical dataset images, and the next is to construct the related models for the classification of the given medical images (Litjens et al., 2017).

In terms of a large medical dataset, the diagnosis becomes more accurate. Thus, most of the diagnostic techniques are systematic with the image classification approaches. Still, these tasks have been accomplished with greater significance. Medical classification relies on the shape, texture, and color of the given medical images. These features are used in the medical field with their combinations. Over the past decade, researchers have used these methods with shallow models to classify medical images. The problem in using shallow models is the extracted features of the images; usually, they are low-level features, and for the features that require high-level extraction, the generalization capacity is still low (Livia Faes et al., 2019).

Now, Deep Learning architecture can be used for classifying medical images. Here, Deep Learning-based models are used for creating class labels of medical images with a more efficient method. There are many fusion methods for implementing the high-level features and also the existing feature. In this, we obtain the best performance and the accuracy of the classification of the medical images (Chan et al., 2015).

8.2 DEEP LEARNING ARCHITECTURE AND ITS NEURAL NETWORKS

The human nervous system is the inspiration for the construction of Artificial Neuron Networks (ANNs). Perceptron Neural Network is based on the human brain system, both structurally and conceptually. It contains both the input and output layers and categorizes the patterns with linear separability (Wiens & Shenoy, 2018). The Neural Network consists of interconnected layers, which are input, hidden, and output layers. The general structural design of the Neural Network is shown in Figure 8.1. Each neuron in the network is activated through the activation function with the input layers and provides the output, which gives the subsequent layers of the network (Kim et al., 2017).

When there are numerous hidden layers, the architecture is called Deep Neural architecture, which is more complex. Deep Learning has a variety of applications such as computer vision applications involving object identification, speech, and facial recognition as well as medical imaging. With larger databases and a wider knowledge base with successful intentions, many explorations can be done with the aid of Deep Learning approaches (Litjens et al., 2017; Liu et al., 2019).

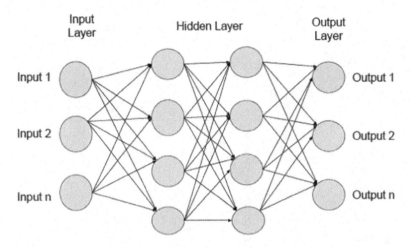

FIGURE 8.1 Neural network architecture.

8.3 CNN-BASED MEDICAL IMAGE CLASSIFICATION

Medical images are classified for the prediction and diagnosis of the patients based on the requirement of the medical analysis. Including many platforms, Deep Learning models are used to diagnose patients in clinical healthcare applications. In these healthcare applications, the most important thing is the patient's health, in order to predict and diagnose a disease (Huang et al., 2019).

With the proper treatments according to the severity of the disease, the clinical outcome will be good healthcare policies. Deep Learning models have been used by many researchers for implementing a wide variety of clinical healthcare tasks. There are publicly available datasets that can be used for the classification of medical information for research purposes. Since there are abundant datasets for classification, interest in applying Deep Learning methods to clinical healthcare applications has been increasing (Lakhani & Sundaram, 2017).

Figure 8.2 illustrates the classification framework based on CNNs. In order to achieve the classification of the dataset, the dataset first needs to be preprocessed; then, the data is selected; the data is cleaned; and, finally, the feature extraction method is used (Yadav & Jadhav, 2019). A CNN-based method has been used broadly in medical classification systems. When processing on small datasets, a CNN has the capability of classification with a higher performance. To improve the performance of using CNN methods, there are a few different strategies, including data augmentation and transfer learning (Sarraf & Tofighi, 2016). SVM classifiers are also used in the transfer learning of VGG-16 and Inception-V3 because SVMs also provide efficient performance. Using data augmentation prevents the over-fitting of the network model.

A CNN is an admirable feature extractor, so it is used in the classification of medical images, which evade the hurdle and the expensive feature extraction. Dataset classification using CNNs provides a greater degree of accuracy and sensitivity for the datasets. Therefore, utilizing transfer learning with data augmentation prevents the

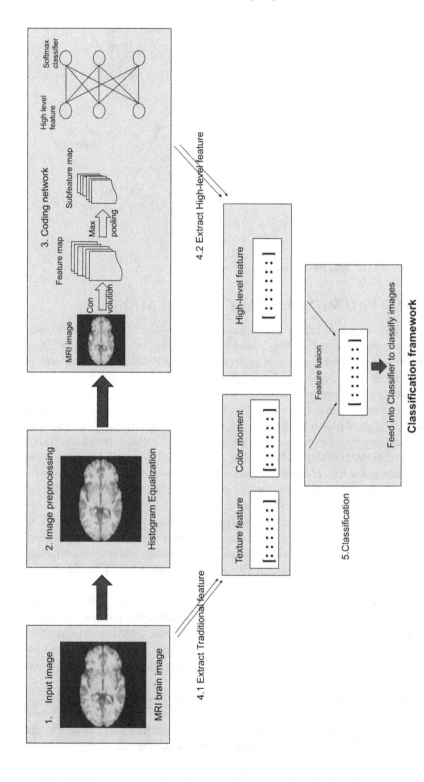

FIGURE 8.2   CNN-based classification framework.

overfitting of the model, which results in a better solution for the classification of medical imaging tasks (Suresh & Mohan, 2019).

8.3.1 Deep Features and Fusion with Multi-Layer Perceptron

Nowadays, image classification is done by computer-based techniques, which have become popular with the rapid improvement in computer technologies. Medical image classification is done through a Computer-Aided Diagnosis (CAD) system, which is a quick and exact annotation of the medical images. In most medical fields, medical images are vital to the procedure of CAD systems. In traditional systems, the features related to the medical images may vary, such that it may focus on different regions, contrast, and also the white balance (De Fauw et al., 2018). Pixel intensity and texture may also be different as they are the inner structures of the medical images. Because of these things, the classification of medical images is challenging. Thus, Deep Learning approaches have become the trending topic in the computer science field and in its applications. Therefore, Deep Learning models are used to solve these image classification problems. The classification is done in an efficient way by constructing an end-to-end method, i.e. Deep Learning models label the classified medical images. In addition to this method, the model may suffer from high computational costs, as it has high-resolution medical images with a smaller number of dataset sizes, which results in restrictions of the models (Wiens & Shenoy, 2018).

To overcome these challenges, Coding Networks with Multi-layer Perceptrons (CNMPs) are incorporated in the Deep Learning models and have been associated with extracting the high-level features within Deep CNNs and other existing methods. With this, categorization is based on the Deep features within the Convolutional Neural Network as a coding network and the outcome is the label for medical images with high-level features classification being based on the supervised manner (Litjens et al., 2017).

Convolution Neural Networks have been extensively used in image classification, image identification, and object recognition, with more accuracy. CNNs usually consist of convolutional layers, pooling layers, fully connected layers, and finally a softmax layer. The convolutional layer and pooling layer are used for the extraction of features of the given medical images; the softmax layer is used as a classifier (Lai & Deng, 2018). The major steps of the Deep Learning models for the image classification are as follows:

1. Image preprocessing.
2. Identifying the activation function for the Neural Network.
3. Initializing the weights as the initial weights are important for a Deep Network to learn.
4. Augmenting the data, which is the most important step for medical image classification.
5. Regularizing, so that the dropout can also be used for overfitting. This trains the models to protect from underfitting and overfitting.
6. Choosing the learning rate.
7. Designing models, based on Deep Learning architecture, for image classification.

8.3.2   ECG ARRHYTHMIA CLASSIFICATIONS

Deep Learning methods have been implemented for ECG signals for medical image classification using CNN, which is most suitable for feature extraction of the given images. An electrocardiogram (ECG) is a non-stationary physiological signal, which represents the electrical activity of the heart and also measures mental stress. In the case of the prediction and classification for medical purposes, Deep Neural Networks (DNNs) are also used, as they have been proven to be suitable for the classification of medical tasks accordingly. Cardiovascular disease (CVD) remains the biggest cause of human death (Livia Faes et al., 2019). When dealing with a huge amount of heterogeneous data, the traditional method still lacks in achieving accuracy in diagnosis. In addition to this, there is no sufficient equipment or enough medical experts for the particular disease diagnosis. There is a necessity for automatic, reliable, and low-cost monitoring and diagnosis systems, such as the Computer-Aided Diagnosis (CAD) systems. These systems provide a summary of the health condition of patients, which are monitored based on the signals and functionality of the subsequent organ (Chan et al., 2015).

The CAD system helps in providing direct solutions for particular information about a disease. In modern CAD systems, ECG signals have been captured for detecting arrhythmia while monitoring heartbeats, thereby, increasing the accuracy of the prediction and diagnosis. Some challenges regarding the classification of arrhythmia are as follows:

1. While the ECG signal is being captured, the symptoms of the arrhythmia might not be seen.
2. Depending on different factors such as age, gender, physical conditions, and routine, ECG signal properties may vary.
3. Morphology does not remain stationary, as it varies according to the patient's physical state.
4. The higher probability results in a false diagnosis of arrhythmia.
5. Artifacts, noise, and interference may also affect morphology variations in the captured ECG signals.

Heart disease can be detected by the ECG signal. The ECG provides the two types of signal, electrical waves and the strength of the electrical activity. This electrical wave is passed through the system of the heart, which shows whether it is regular or irregular. Then, the strength of the electrical activity shows if any parts of the heart are bulky or overworked. Therefore, any disorders in the electrical activity of the heart neural cells that affect the ECG signals are denoted as arrhythmia (Shen et al., 2017). Some performance measurements signify the accuracy of the classifiers. The classification may result in normal or abnormal, according to the prediction, such that,

1. True Positive (TP): Positive class prediction correct.
2. True Negative (TN): Negative class prediction correct.

3. False Positive (FP): Positive class prediction incorrect.
4. False Negative (FN): Negative class prediction incorrect.

Therefore, the accuracy of these classes is calculated as follows:

$$Accuracy = \frac{TP + TN}{TP + TN + EP + FN}$$

Deep Learning approaches are also used for the extraction of features that are given to the classifiers for productive information. CNNs are broadly used for classification methods with the supervised learning approaches, as well as for feature extraction and noise filtering. A CNN model is also used for the diagnosis of arrhythmia as it shows greater performance than previous methods (Ebrahimi et al., 2020).

8.3.3 Classification of Tuberculosis-Related Chest X-Ray

In developing countries, tuberculosis (TB) is a major health issue. Many people have been infected with TB, which mainly affects the lungs and occasionally other parts of the body. A chest x-ray (CXR) is the basic tool for TB diagnosis; there are also a few other procedures that are used for detecting this disease. The screening for TB is done accordingly and Deep Convolutional Neural Network (DCNN) has become the best technique for TB surveillance and detection (Sathitratanacheewin et al., 2020). Here, a supervised learning algorithm is used for training the data that are used for identification and prediction. The color pixel values of the CXR images are used for the detection of TB and labels are provided to detect whether it is normal or infected. With the data provided, the algorithm can be further developed so as to detect the disease in its early stages (Liu et al., 2019).

In the most common Deep Learning models, the results may differ, such that a CXR image with fewer lung lesions may be labeled as normal; so, this has to be overcome using computer-aided detection for tuberculosis. In the supervised learning model, the requirement and disease severity level may also result in a diagnosis. Thus, the DCNN model is used for TB detection and further for classification accordingly. For the classification of TB with the chest x-ray, DCNN is used for classifying the normal and other CXR. Therefore, the accuracy may improve using high-resolution medical images so as to more easily detect and diagnose for the classification of the disease (Wang et al., 2017).

8.3.4 Clinical Image Classification of Infectious Keratitis

The tenderness and damage of corneal tissues that are from the pathogen which develops in the cornea is the most general form of corneal disease, known as infectious keratitis. This type of disease must be treated immediately to minimize the damage to the cornea. For the classification of the clinical images, the sequential-level deep model is used to differentiate between corneal diseases. Usually, ophthalmologists treated this disease based on data provided and information that are inclusive. Now, these diseases are treated with Deep Learning methods as it is widely used in many medical applications (Litjens et al., 2017).

The visual feature of the infectious lesion in the cornea is recognized as infectious keratitis. In a computer-aided diagnosis, Deep Learning models have been broadly used in image classification and image recognition. Usually, AI algorithms are used in medical image classification, also used to detect retinal diseases that must be diagnosed correctly. There is some classification in the datasets, according to the infection of the corneal, such as bacterial keratitis (BK), fungal keratitis (FK), and herpes simplex virus stromal keratitis (HSK). The images that are collected from the patients are to be annotated based on this, and the disease is identified. The labeling may be of four types: infectious lesion, other than infected part, conjunctival injection, and anterior chamber (Shen et al., 2017).

By the classification of these parts, the implementation is done for the diagnosis of the clinical image. In image-based classification, ophthalmologists must follow some procedures: initially, a diagnosis image of the individual patients is collected and labeled with BK, FK, or HSK; then, the structured information that is related to the patient record must be briefly identified and diagnosed. Visual awareness should play a vital role as it illustrates the relation involving shape, texture, color, size, and correlation. When using the image annotated data, they require a huge quantity, so an AI-based diagnostic system is used to rectify this issue. Leading to higher diagnostic accuracy, AI helps clinicians in their improvement. In addition to the sequential learning model, multi-modal learning systems are also used to increase precision (Xu et al., 2020).

8.3.5 DIABETIC RETINOPATHY

Diabetic retinopathy (DR) usually happens when there is a high blood sugar level in diabetic patients. It is identified as Type-1 and Type-2 diabetes. Type-1 diabetes occurs due to insulin not being produced by the pancreas, whereas Type-2 diabetes occurs due to insulin not correctly reacting in the body tissues, which leads to high blood sugar (Saxena et al., 2020). DR is generally related to eye disease, which results in permanent blindness when it is not treated properly at the early stages. DR can be considered as a symptom of disorderly diabetes, but it is related to diabetes and not commonly incorporated. An automated system is used for screening the patient periodically, which must find DR in the initial stage. For the automated detection of this disease, CNN emerged to be ideal in dealing with medical images (Gulshan et al., 2016).

DR images can classify among the different datasets, and these images are from annotation with camera light effects. CNN must detect and classify these images, where there is no need for unrelated data. This data must be removed to improve performance. The images usually contain a black border, rotation of the eye, imposture, and so on; these must be eliminated. The pre-processing methods and data augmentation are used for the classification of related medical images (Krause et al., 2018).

The automated system identified DR with the colour images of the retina to identify this at an early stage in order to control the vision loss. For a nearly definite function, the classifier models have to be built accordingly. These classifiers are made to detect automatically for the early stages of DR. This enhances the classifier to detect the DR as early as possible to limit vision loss. By improving the classifier, it results in achieving greater performance; therefore, the classifier has to be trained accordingly (Ting et al., 2017).

8.3.6 TUMOR STAGE CLASSIFICATION OF PULMONARY LUNG NODULES

When compared with other types of cancers, lung cancer is one of the foremost diseases worldwide that leads to high death rates. As such, the early detection of this disease may increase survival rates. Computed tomography (CT) images are used for identifying the pulmonary nodule boundaries and this mission is very critical to the computer-aided systems (Huang et al., 2019). A DCNN has been introduced for the automatic classification of the pulmonary lung nodules. According to the samples of these nodules, they are classified as non-cancerous, benign, or malignant samples. DCNNs are trained for these classifications of the disease. Some classification techniques used by the existing systems such as textures, density, and statistical are complex and time-consuming. Nowadays, without considering the manually extracted features, several researchers have developed architectures that can be learned and trained. Therefore, with the help of these techniques, the detection and classification of these medical images can be automatic (Lakshmanaprabu et al., 2019).

Usually, the lung nodule system has challenges to overcome, such as segmentation, extraction, and classification. A diagnosis is based on both the existing systems and the automated systems. Thus, the quality of the images is improved using quality thresholding. Based on the region growing methods, the segmentation is done.

Then, the feature extraction such as shape, the texture is to be extracted using various AI algorithms, and, finally, the classification is done using SVM classifiers. In the initial stage of detection, the preprocessing of the images plays an essential part. Accuracy can be improved by increasing the number of layers in the CNN to enhance efficiency (Suresh & Mohan, 2019).

8.3.7 CLASSIFICATION OF ALZHEIMER'S DISEASE

Magnetic resonance imaging (MRI) scans are used for the diagnosis of Alzheimer's disease (AD) with the help of Deep Learning algorithms. For an individual diagnosis of AD, a CNN can be a vital tool in the classification of this disease. For maintaining and assisting the patients, a CNN provides the structural MRI for schedule practice. The diagnostic performance may also be increased using biomarkers. A structural MRI is used by the learning methods with the computer-aided classification of AD and the most familiar method is the Support Vector Machine (SVM) (Sarraf & Tofighi, 2016). The clinical diagnosis can be achieved by facilitating the classification model, in which the SVM gathers features and information provided by the MRI images (Basaiaa et al., 2019). However, these machine learning methods are computationally complex and time-consuming. Instead of these methods, Deep Learning methods perform better in many fields such as computer vision, recognition systems, segmentation, and so on.

In addition to the single cross-sectional brain structural MRI scan, Deep Learning methods and algorithms can be implemented to predict and classify medical images to diagnose patients. When Deep Learning Neural Networks are used, the artificial neurons are arranged such that the neurons are fully connected in every layer of the network. A CNN is a specific type of feed-forward network, which is used to process biologically inspired medical images. A CNN Deep Learning network works as follows:

a. The Model with the data input, which learns from the inputs.
b. Throughout the input to output, the network is propagated with the information.
c. The error signal calculation (the difference between the output and the target value).
d. Error signals have to be propagated. Weights have to be adjusted accordingly, repeat the above steps as the error becomes relatively small. At last, the trained network has to predict the class and classifiers are used to separate the new classes.

Now a CNN has the ability to classify the raw images and has achieved accuracy measures with true positives and true negatives having been recognized. CNNs have been proven to be the model for the automation of early detection of AD, which is helpful for organizing and managing the patients' needs (Weiner et al., 2017).

8.3.8 CLASSIFICATION OF PRIMARY BONE TUMORS ON RADIOGRAPHS

Deep Learning models can be used to classify bone tumors and have been compared to the performance of radiologists' ability to detect tumors from radiographs. Tumor classification is labeled benign, intermediate, or malignant. Diagnosis of a tumor usually follows up the reports and age of the patients, and the tumor's appearance such as margin, location, and size of the demographic information (Bradshaw et al., 2018). The classification depends on the behavior and requirements of histology management. Usually, tumors are classified in the following steps:

i. Preprocessing
ii. Training and inference
iii. Model evaluation
iv. Statistical analysis

By using Deep Learning models, classifying tumors can be correct and precise. Radiologists perform manual cropping for labeling the lesions; the three-way classification and binary discriminatory were designed for enhancing the performance. Classification must be done properly in order to prescribe for patients accurate treatments; otherwise, it can lead to improper treatment with needless biopsies (Lakhani & Sundaram, 2017). Thus, the records should be appropriate and arranged accordingly. There are some centers that have databases of patients diagnosed with a bone tumor, with or without a biopsy and pathology. The automated computer-aided systems are used to classify primary bone tumors on the radiographs using the many databases available, in order to increase both the accuracy and the performance (Yu Hea et al., 2020).

8.3.9 CLASSIFICATION OF BRAIN TUMORS

Atypical and uncontrolled progress in the synapses of the brain is called brain cancer (tumor). Based on human competence, the individual head is rigid and also volume constraints; tumors may also distend into other parts of the body, which manipulate individual abilities. The World Health Organization (WHO) recognizes two significant kinds of brain tumors (BT); if a tumor is expanding within the brain, it is called a Primary Brain Tumor, and if a tumor distends into the brain from other organs of the body, it is referred to as a Secondary Brain Tumor (Mehrotra et al., 2020). Therefore, the classification of tumors is done for further diagnosis and treatment. Magnetic Resonance Imaging (MRI) is used for these types of assignments because the quality of the image does not depend on ionizing radiation. Artificial intelligence (AI), in the form of Deep Learning, is used in medical imaging for segregating and classifying brain tumors (Ting et al., 2017).

Many Deep Learning algorithms are also implemented for identifying and matching tumors related to the training images of the Neural Networks. This is also enhanced for identifying tumor location, size, and type for the proper treatment of the patients. For inspecting tumors, Deep Learning (DL) makes the selection with the functioning of the brain in the data processing and creates many suitable protocols (Yu Hea et al., 2020). Thus, layers with non-linear features are arranged in well-mannered ways for extracting features of the images. A CNN is generally used for these purposes, to organize and classify the medical images with feature extractions. First, the MRI brain image dataset is given as input, the preprocessing is done there to extract features of the images with the image labels. Then the image is augmented for the segregation of images with predefined labels, if any were available. The Deep Neural Network must be trained and tested with image datasets, which gives the results as network output. Finally, the tumor is classified as Benign or Malignant (Lai & Deng, 2018).

There are many optimization techniques used for classification, including Support Vector Machine, because they improve accuracy. The main approach of this classification is the well-built DL methods for detecting and processing medical images. Then, figure out which DL method is most useful for a particular problem and, at last, evaluate the performance with the fine-tuned systems. A CNN is utilized for multimodal dimension mappings after yielding ideal training outcomes (Zhu et al., 2020).

8.3.10 CLASSIFICATION METHODS FOR DIAGNOSIS OF SKIN CANCER

Lately, Deep Learning methods have been used for diagnosing skin cancer. There is not much awareness among people, in addition to its requiring adequate facilities for identifying and treatment. There are commonly available datasets to differentiate between benign lesions and malignant lesions in a variety of images such as clinical, dermoscopic, and also histopathological images (Bradshaw et al., 2018). With widespread ideas of using AI applications in the medical field, this AI system achieves a higher performance in accuracy and in terms of classification. Non-melanoma skin cancer occurs in both genders. Detecting the disease in the early stages guarantees survival. Computer-aided diagnostic (CAD) systems are widely used across medical fields for diagnosing diseases (Wiens & Shenoy, 2018).

Numerous AI algorithms are used in medicine in order to help clinical care improve its performance. Thus, there are three approaches for detecting skin lesions, including:

    i. Clinical images
   ii. Dermoscopic images
  iii. Histopathology images

Here, a diagnosis is formed by AI algorithms with these major types of image skin lesions. The schedule captures images of various positions and sizes to examine and integrate into the patients' medical records. Next with the dermoscopic images, the dermatopathologists identify using the biopsy tissues (Goyal et al., 2020). DL is most popular in the field of histopathologic images with the sliding of the images.; these samples are used for detecting various types of cancers, such as skin, lung, and breast. Based on the classification of images, the colour, texture, size, and appearance of the skin lesions decide the inter-class and intra-class distinction (Mike et al., 2019).

8.3.11 COVID-19 Detection in CT Images

The virus COVID-19, having originated from SARS-CoV-2, is formally acknowledged as a pandemic by the World Health Organization (WHO). COVID-19 is tremendously transmittable and has an effect on the fatal acute respiratory distress syndrome (ARDS). Controlling the COVID-19 spread in the early stages of detection and analysis is a significant issue. Reverse-transcription polymerase chain reaction (RT-PCR) testing is the most wide-ranging method for transmission. Low sensitivity may occur in the early stages of screening. Chest scans, including x-rays and CT scans, are used for the detection of COVID-19 according to the morphological pattern of lung lesions. These days, Deep Learning is used in all fields, especially in the medical field. Thus, the screening of CT images for the detection of COVID-19 is being done by many Deep Learning methods (Abbas et al., 2020). To overcome the problem of slicing the CT images, the Deep Learning process is automated for the training and testing of images. The screening of COVID-19 is done by voting-based methods. Therefore, the efficient-based method is used for screening purposes; the datasets that are associated with CT-based analysis of the patients.

A computed tomography scan, or CT scan, gives the complete structure of blood vessels, soft tissues, bones, and organs. Thus, the CT images allow physicians to classify the internal structures such as size, texture, and density. CT scans also help in slicing a set of an organ of the particular region of the body without overlapping the other body structures. Thus, CT scans provide a much more meticulous depiction of the patient's state than conventional x-rays. With this accurate information about the patient's condition, the exact medical problem can be detected and treated. Hence, using several Deep Learning-based methodologies, COVID-19 screening in CT scans is proposed. The vital methods are: (i) a proficient Deep Learning model for the transmission of COVID-19 in CT scans and (ii) to deal with the aforementioned questions concerning the two biggest datasets, and (iii) a voting-based assessment approach. The prospects of Deep Learning models for the task of COVID-19 detection on CT images have been described by Silva et al. (2020).

8.3.12 MRI Harmonization and Confound Removal Using Neuro-Imaging Datasets

Counting numerous exceptionally multi-site multi-scanner datasets, huge x-ray neuro-imaging datasets are accessible. Grouping and division assignments with two distinctive network structures are completed using Deep Learning models. In view of Deep Learning models, other generative models have been utilized to fit x-rays. To approve the created "fit" images, the danger of obscure mistakes spreading through pipelines and influencing the aftereffects of any finished examination. Area Ill-disposed Neural Networks (DANNs) powerful space move to be accomplished, forecasts should be founded on the component. Class names are an indicator that predicts, at that point to foresee the wellspring of the information from the space classifier. To blend for the scanner, the structure can be used to eliminate other jumbles. An area variation strategy can be adjusted to orchestrate x-ray information for a given errand using CNNs, by making an element space that is invariant to the obtaining scanner. The expulsion of scanner data from the component space keeps the data from driving the expectation. Another strategy for an ill-disposed area transformation was presented, in which, instead of utilizing an inclination inversion layer to refresh the space classifier contrary to the assignment, they utilized an iterative preparing plan. Further, we will think about a scope of information likely situations, for example, the impact of having restricted measures of preparing information, and the impact of having various dissemination of information for various scanners, and show that the preparation system is just adjusting to managing these extra difficulties (Henschel et al., 2020).

The space classifier must be used to determine how well area data are being eliminated, in the event that it is able to precisely foresee the space prior to forgetting. The span of a clump increment is not the multiple times that may be normal, with the increment from one-forward and in a reverse pass to three frontward and reverse passes per age, because the second two trouble works just update subsets of the boundaries, lessening the number of computations that should be finished. The various methodologies are not easily analyzed, particularly for the division assignment, and there is a spotlight on various phases of the pipeline. In any event, for the division task, where there is no huge upgrading in the presentation, the forgetting guarantees us that the yield esteems are not motivated by the scanner.

This might be significant for downstream investigation, for example, contrasting white matter volumes, permitting the information to be mutual across destinations. The methodology can be expanded to allow us to eliminate extra frustrates, and have shown a way that this could likewise be expanded to allow for eliminating ceaseless jumbles, such as age. For each perplex to be taken out, we involve two extra misfortune capacities and two extra frontward and in reverse passes. Thus, the preparation time resolves increment with each extra frustration, introducing an expected limit (Dinsdale et al., 2020).

8.3.13 DEEP LEARNING IN SPATIOTEMPORAL CARDIAC IMAGING

In medical image studies, spatial and temporal data are widely used with the help of AI and Deep Learning. There are many models of cardiac imaging, such as MRI, CT, Ultrasound, and others. Using these modalities, the classification and detection of heart diseases are identified with the visualization effects of the structures. While handling the classification and segmentation of the cardiovascular anatomy, cardiovascular imaging is mainly focused on AI and Deep Learning-based strategies. Also, the imaging modalities are based on the morphologies in the Left/Right Ventricle (LV/RV), the coronary tree, and the aorta, the valve plane. In Cardiac Magnetic Resonance Imaging (CMRI), Deep and Machine Learning algorithms can be used for the identification and detection of cardiac functions and for the accuracy of ventricular volumes. In echocardiography (ECHO), the standard views are used for the detection and the segmentation of the LV (Bello et al., 2019).

All the Deep Learning-based techniques that are used are based on an initial learning rate, batch size, optimization algorithm, training and testing data, loss function, and model parameters. Based on the researchers' data, the preprocessing, datasets, clinical analysis, learning method, data annotations, training approach, and test accuracy are all helpful in estimating performance and accuracy. Using a CNN as a classifier, the preprocessing step is done by the Feature Asymmetry for the classification of cardiac action. Thus, a CNN is used for the prediction of the visualization of planes and the orientation. Many methods have been used to improve the detection of a cardiac phase that leads to diagnosing cardiac-related issues. Some limitations need to be considered including computational complexity and the training of models, which may lead to high computational costs. In Deep Learning methodologies, some parameters, such as robustness, have been used, and the performance of the clinical data diagnoses the cardiac symptoms based on the framework. According to the application of Deep Learning, the performance may vary (Hernandez et al., 2021).

8.3.14 LIVER TUMOR CLASSIFICATION USING DEEP LEARNING MODEL

These days, clinical indicative sweeps, for instance, Computed Tomography (CT) and Magnetic Reverberation Imaging (MRI) are significant for the analysis and evaluation of treatment for numerous illnesses. Liver tumor (malignancy) is the most well-known tumor illness around the world, and it prompts critical fatalities on a yearly basis. Exact tumor estimations (from MRI and CT), as well as tumor size, area, and shape, can help specialists in making exact disease evaluation and treatment arrangements. The programmed division of liver and tumor faces numerous difficulties including the difference level among liver and tumor is moderately little, there are fluctuating sizes and kinds of liver tumors and abnormalities in tissues. Most Deep Learning analysts accept that the more profound model is the best. By and by, the more profound model faces popping/disappearing slope issues, which impede combination through preparing (Moghbel et al., 2018).

DenseNets give an immediate association among the whole layers, which improves the progression of slopes and data through all the layers. Also, thick associations decline the overfitting issue when chipping away at a more modest preparing dataset. DenseNets displayed great exhibitions in image classification. Accordingly, numerous specialists expand the use of DenseNets to division issues. The engineering tends to substantial calculation and memory utilization issues where neighborhood and worldwide highlights are separated and intertwined for exact liver and tumor division. High memory use limits input size, network profundity, and channel size, which are significant factors in accomplishing elite. Also, the high computational expense of 3-D convolutions limits preparation for an enormous scope dataset (Nasser Alalwan et al., 2021).

8.4 CONCLUSION

Recently, Deep Learning has become the most suitable for computer vision techniques, particularly in the medical imaging field. Deep Learning has the potential to renovate healthcare; yet, extensive expertise is vital to train such models. In medical imaging, the diagnosis with precision and speed is done through artificial intelligence (AI) including Deep Learning. Thus, most of the diagnostic technique is systematic, as it has intelligent classification approaches with computer-aided decision support systems. With larger databases and a wider expanse of knowledge with successful intentions, much exploration can be done with the assistance of Deep Learning methods. A CNN is an admirable feature extractor, as such, it is used in the classification of medical images, which evades the hurdle of expensive feature extraction. Comparing dataset classification using a CNN provides greater accuracy and sensitivity on their datasets. Therefore, using transfer learning with data augmentation prevents the overfitting of the model, which results in a better solution for the classification of medical imaging responsibilities.

REFERENCES

Abbas, Abdelsamea, M., & Gaber, M. (2020). Classification of covid-19 in chest x-ray images using DeTraC deep convolutional neural network. *medRxiv*. doi:10.1101/202 0.03.30.20047456

Alalwan, N., Abozeid, A., ElHabshy A. A., & Alzahrani, A. (2021). Efficient 3D deep learning model for medical image semantic segmentation. *Alexandria Engineering Journal, 60*, 1231–1239.

Basaiaa, S., Agostaa, F., Wagnerc, L., Canua, E., Magnanib, G., Santangelob, R., & Filippi, M. (2019). Automated classification of Alzheimer's disease and mild cognitive impairment using a single MRI and deep neural networks. *NeuroImage: Clinical, 21*, 101645. doi:10.1016/j.nicl.2018.101645

Begoli, E., Bhattacharya, T., & Kusnezov, D. (2019). The need for uncertainty quantification in machine-assisted medical decision making. *Nature Machine Intelligence, 1*(1), 20–23.

Bello, G. A., Dawes, T. J. W., Duan, J., Biffi, C., Gibbs, R., de Marvao, A., Howard, L. S. G. E., Simon, J., Wilkins, M. R., Cook, S. A., Rueckert, D., & O'Regan, D. P. (2019). Deep learning cardiac motion analysis for human survival prediction. *Nature Machine Intelligence, 1*, 95–104. doi:10.1038/s42256-019-0019-2

Bradshaw, T., Perk, T., Chen, S., Im, H.-J., Cho, S, Perlman, S., & Jeraj, R. (2018). Deep learning for classification of benign and malignant bone lesions in [F-18]NaF PET/CT images. *The Journal of Nuclear Medicine, 59*, 327.

Chan, T., Jia, K., Gao, S., Lu, J., Zeng, Z., & Ma, Y. (2015). PCANet: A simple deep learning baseline for image classification. *IEEE Transactions on Image Processing, 24*(12), 5017–5032.

De Fauw, J., Ledsam, J. R., Romera-Paredes, B., Nikolov, S., Tomasev, N., Blackwell, S., Askham, H., Glorot, X., O'Donoghue, B., Visentin, D., van den Driessche, G., Lakshminarayanan, B., Meyer, C., Mackinder, F., Bouton, S., Ayoub, K., Chopra, R., King, D., Karthikesalingam, A., Hughes, C. O., Raine, R., Hughes, J., Sim, D. A., Egan, C., Tufail, A., Montgomery, H., Hassabis, D., Rees, G., Back, T., Khaw, P. T., Suleyman, M., Cornebise, J., Keane, P. A., & Ronneberger, O. (2018). Clinically applicable deep learning for diagnosis and referral in retinal disease. *Nature Medicine, 24*(9), 1342–1350. doi:10.1038/s41591-018-0107-6.

Dinsdale, N. K., Jenkinson, M., & Namburete, A. I. L. (2020). Deep learning-based un-learning of dataset bias for MRI harmonisation and confound removal. *NeuroImage*. doi:10.1016/j.neuroimage.2020.117689

Ebrahimi, Z., Loni, M., Daneshtalab, M., & Gharehbaghi, A. (2020). A review on deep learning methods for ECG arrhythmia classification. *Expert Systems with Applications*. doi:10.1016/j.eswax.2020

Faes, L., Wagner, S. K., Jack Fu, D., Liu, X., Korot, E., Ledsam, J. R., & Anil, P. (2019). Automated deep learning design for medical image classification by health-care professionals with no coding experience: A feasibility study. *Lancet Digital Health, 1*, e232–e242.

Goyal, M., Knackstedt, T., Yan, S., & Hassanpour, S. (2020). Artificial intelligence-based image classification methods for diagnosis of skin cancer: Challenges and opportunities. *Computers in Biology and Medicine, 127*, 104065. doi:10.1016/j.compbiomed.2020.104065

Gulshan, V., Peng, L., Coram, M., Stumpe, M. C., Wu, D., Narayanaswamy, A., Venugopalan, S., Widner, K., Madams, T., Cuadros, J., Kim, R., Raman, R., Nelson, P. C., Mega, J. L., & Webster, D. R. (2016). Development and validation of a deep learning algorithm for detection of diabetic retinopathy in retinal fundus photographs. *JAMA, 316*(22), 2402–2410. doi:10.1001/jama.2016.17216.

Hea, Y., Panb, I., Baoa, B., Halseyb, K., Changc, M., Liua, H., & Bai, H. X. (2020). Deep learning-based classification of primary bone tumors on radiographs: A preliminary study. *EBioMedicine, 62*, 103121. doi:10.1016/j.ebiom.2020.103121

Henschel, L., Conjetia, S., Estrada, S., Diers, K., Fischl, B., & Reuter, M. (2020). Fastsurfer – A fast and accurate deep learning based neuroimaging pipeline. *Neuroimage, 219*(10).

Hernandez, K. A. L., Rienmüller, T., Baumgartner, D., & Baumgartner, C. (2021). Deep learning in spatiotemporal cardiac imaging: A review of methodologies and clinical usability. *Computers in Biology and Medicine, 130*, 104200. doi:10.1016/j.compbiomed.2020.104200

Huang, X., Sun, W., Tseng, T.-L. B., Li, C., & Qian, W. (2019). Fast and fully-automated detection and segmentation of pulmonary nodules in thoracic CT scans using deep convolutional neural networks. *Computerized Medical Imaging and Graphics, 74*, 25–36. doi:10.1016/j.compmedimag.2019.02.003

Kim, S. J., Cho, K. J., & Oh, S. (2017). Development of machine learning models for diagnosis of glaucoma. *PLoS One, 12*(5), e177726.

Krause, J., et al. (2018). Grader variability and the importance of reference standards for evaluating machine learning models for diabetic retinopathy. *Ophthalmology, 125*(8), 1264–1272.

Lai, Z., & Deng, H. (2018). Medical image classification based on deep features extracted by deep model and statistic feature fusion with multilayer perceptron. *Computational Intelligence and Neuroscience, 13*. doi:10.1155/2018/2061516

Lakhani, P., & Sundaram, B. (2017). Deep learning at chest radiography: Automated classification of pulmonary tuberculosis by using convolutional neural networks. *Radiology, 284*, 574–582.

Lakshmanaprabu, S., Mohanty, S. N., Shankar, K., Arunkumar, N., & Ramirez, G. (2019). Optimal deep learning model for classification of lung cancer on CT images. *Future Generation Computer Systems, 92*, 374–382.

Litjens, G., Kooi, T., Bejnordi, B. E., Setio, A. A. A., Ciompi, F., Ghafoorian, M., van der Laak, J. A. W. M., van Ginneken, B., & Sánchez, C. I. (2017). A survey on deep learning in medical in medical image analysis. *Medical Image Analysis, 42*(9), 60–88.

Liu, X., Faes, L., Kale, A. U., Wagner, S. K., Fu, D. J., Bruynseels, A., Mahendiran, T., Moraes, G., Shamdas, M., Kern, C., Ledsam, J. R., Schmid, M. K., Balaskas, K., Topol, E. J., Bachmann, L. M., Keane, P. A., & Denniston, A. K. (2019). A comparison of deep learning performance against health-care professionals in detecting diseases from medical imaging: A systematic review and meta-analysis. *The Lancet Digital Health, 1*, e271–e297. doi:10.1016/s2589-7500(19)30123-2.

Mehrotra, R., Ansari, M. A., Agrawal R., & Anand, R. S. (2020). A transfer learning approach for AI-based classification of brain tumors. *Machine Learning with Applications, 2*, 100003. doi:10.1016/j.mlwa.2020.100003

Mike, V., & Kajsa, M., & Ailo, B. L. (2019). Reproduction study using public data of: Development and validation of a deep learning algorithm for detection of diabetic retinopathy in retinal fundus photographs. *PloS One, 14*(6).

Moghbel, M., et al. (2018). Review of liver segmentation and computer assisted detection/diagnosis methods in computed tomography. *Artificial Intelligence Review, 50*(4), 497–537.

Sarraf, S., & Tofighi, G. (2016). DeepAD: Alzheimer's disease classification via deep convolutional neural networks using MRI and fMRI. Retrieved from https://www.biorxiv.org/content/early/2016/08/21/070441

Sathitratanacheewin, S., Sunanta, P., & Pongpirul, K. (2020). Deep learning for automated classification of tuberculosis-related chest X-Ray: Dataset distribution shift limit diagnostic performance generalizability. *ScienceDirect, Heliyon, 6*, e04614. doi:10.1016/j.heliyon.2020.e04614

Saxena, G., Verma, D. K., Paraye, A., Rajan, A., & Rawat, A. (2020). Improved and robust deep learning agent for preliminary detection of diabetic retinopathy using public datasets. *Intelligence-Based Medicine, 3–4*, 100022. doi:10.1016/j.ibmed.2020.100022

Shen, D., Wu, G., & Suk, H. I. (2017). Deep learning in medical image analysis. *Annual Review of Biomedical Engineering, 19*, 221–248.

Silva, P., Luz, E., Silva, G., Moreira, G., Silva, G., Lucio, D., & Menotti, D. (2020). COVID-19 detection in CT images with deep learning: A voting-based scheme and cross-datasets analysis. *Informatics in Medicine Unlocked, 20*, 100427.

Suresh, S., & Mohan, S. (2019). NROI based feature learning for automated tumor stage classification of pulmonary lung nodules using deep convolutional neural networks. *Computer and Information Sciences*. doi:10.1016/j.jksuci.2019.11.013

Ting, D. S. W., Cheung, C. Y., Lim, G., Tan, G. S. W., Quang, N. D., Gan, A., Hamzah, H., Garcia-Franco, R., San Yeo, I. Y., Lee, S. Y., Wong, E. Y. M., Sabanayagam, C., Baskaran, M., Ibrahim, F., Tan, N. C., Finkelstein, E. A., Lamoureux, E. L., Wong, I. Y., Bressler, N. M., Sivaprasad, S., Varma, R., Jonas, J. B., He, M. G., Cheng, C. Y., Cheung, G. C. M., Aung, T., Hsu, W., Lee, M. L., & Wong, T. Y. (2017). Development and validation of a deep learning system for diabetic retinopathy and related eye diseases using retinal images from multiethnic populations with diabetes. *JAMA*, *318*(22), 2211–2223. doi:10.1001/jama.2017.18152

Wang, X., Peng, Y., Lu, L., Lu, Z., Bagheri, M., & Summers, R. M. (2017). ChestX-ray8: Hospital-scale chest X-ray database and benchmarks on weakly-supervised classification and localization of common thorax diseases (pp. 2097–2106). *arXiv:1705.02315*. doi:10.1109/CVPR.2017.369

Weiner, M. W., Veitch, D. P., Aisen, P. S., Beckett, L. A., Cairns, N. J., Green, R. C., & Trojanowski, J. Q. (2017). Recent publications from the Alzheimer's disease neuroimaging initiative: Reviewing progress toward improved AD clinical trials. *Alzheimer's & Dementia*, *13*, e1–e85.

Wiens, J., & Shenoy, E. S. (2018). Machine learning for healthcare: On the verge of a major shift in healthcare epidemiology. *Clinical Infectious Diseases*, *66*(1), 149–153.

Xu, Y., Kong, M., Xie, W., Duan, R., Fang, Z., Lin, Y., & Yao, Y.-F. (2020). Deep sequential feature learning in clinical image classification of infectious keratitis. *Engineering*. doi:10.1016/j.eng.2020.04.012

Yadav, S. S., & Jadhav, S. M. (2019). Deep convolutional neural network-based medical image classification for disease diagnosis. *Journal of Bigdata*, *6*, 113. doi:10.1186/s4 0537-019-0276-2

Zhu, Y., Gao, T., Fan, L., Huang, S., Edmonds, M., Liu, H., Gao, F., Zhang, C., Qi, S., Wu, Y. N., Tenenbaum, J. B., & Zhu, S.-C. (2020). Dark, beyond deep: A paradigm shift to cognitive AI with humanlike common sense. *Engineering*, *6*(3), 310–345.

9 Deep Learning in Medical Image Classification

A. Suganya and S. L. Aarthy

CONTENTS

DOI: 10.1201/9781003038450-9

9.1   INTRODUCTION

Deep Learning has become a well-known research field over the last several years. It is a kind of Machine Learning, which is an offshoot of Artificial Intelligence. It can possibly rebuild health diagnoses and management by performing categorization challenges for human specialists and by quickly inspecting a tremendous amount of images. Deep Learning, otherwise called Deep Neural Learning, provides a proficient all-the-time learning system that can figure out categorization labels from raw clinical image pixels. As of late, huge advances have been seen by Deep Learning contrasted with other AI (ML) procedures that give computerized finding arrangements; in any case, customary ML strategies are not fitting to deal with complex issues.

Reduced medical services information is no more. It is challenging and exciting for analyzing images as the data are very large with the tremendous headway in clinical imaging gadgets. Because of this quick development in clinical images and strategies, broad and tedious endeavours are needed by a clinical master, open to human error, and may have significant contrasts across the various trained professionals. Programmed infection conclusion by AI strategies gives a substitute arrangement; yet, customary AI techniques are not suitable to manage a perplexing issue. Deep Learning can assist with separating highlights and with making new ones. Additionally, it can find an infection as well as measure the prognostic objective and convey useable clinical predictive representations to support doctors skillfully.

Artificial Intelligence (AI) and Machine Learning (ML) have made quick headway recently in clinical fields such as PC-supported diagnosis, clinical image preparation, image translation, image-guided treatment, image consolidation, image enrolling, image detachment, image recovery, and image assessment.

ML methods address compelling and productive information by mining highlights from the clinical images. These techniques simplify and uphold specialists in diagnosing and anticipating the danger of disorders and cure them easily. They also improve the abilities of clinicians and researchers to perceive how to assess the essential adjustments to disease (Dey et al., 2017). The often-used forms of ML calculations include Linear and Logistic Regression, Support Vector Machines, K-Means, kNN, Random Forest, Naïve Bayes, and others. The restrictions of these techniques include handling raw images, tedious and dependence on skilled data, and require more time for fine-tuning the attributes of the data. However, DL strategies like Convolutional Neural Networks (CNN), Recurrent Neural Networks (RNN), Long Short-Term Memory (LSTM), Generative Adversarial Networks

(GAN), Extreme Learning Models (ELM), and others can separate extracts features automatically from raw information. These models attempt to become familiar with a few degrees of speculation, portrayal, and data consequently from an enormous arrangement of images that show the favoured conduct of information. Indeed, even automatic identification of ailments by traditional clinical imaging techniques has been shown substantial accuracies for quite a long time, creative improvements in AI strategies go up in flames in Deep Learning. DL strategies have accomplished confident enhancements in different pattern recognition techniques, such as speech recognition, text recognition, lip-reading, PC-helped diagnosis, face recognition, and drug testing (Razzak et al., 2018).

This chapter aims to deliver a complete analysis of DL-based procedures in medical image classification issues with regard to recent studies and upcoming methodologies. This section presents the basic insights and the ingenious methodologies of Deep Learning in the field of clinical image classification. Moreover, we discuss the hurdles faced by Deep Learning methods for medical imaging and open scientific issues.

9.2   MEDICAL IMAGE CLASSIFICATION

Before discussing medical image classification, let us discuss the overview of medical imaging and its importance in this section. Also, we will analyze the use of Deep Learning techniques over traditional methods.

9.2.1   WHAT IS MEDICAL IMAGING?

Medical imaging is the process of acquiring images of the inner organs or tissues of the human body for medical uses for tracking health, and to detect and cure lesions and disorders. Further, it helps in the creation of an anatomical and physiological database that can be further used to extract relevant information to diagnose diseases and injuries.

Despite current immense developments in the medical field, for many valuable clinical uses, medical imaging can acquire knowledge about the human body. Different forms of medical imaging skills offer contradictory details regarding the location of the body that needs to be examined or treated clinically.

Medical centers and clinics as well as freestanding radiology and pathology facilities use clinical imaging equipment. Fujifilm, GE, Siemens Healthineers, Philips, Toshiba, Hitachi, and Samsung are the significant makers of these clinical imaging gadgets. Zion market research reported that with the headway and expansion in the utilization of clinical imaging, the worldwide market for this equipment made for clinical imaging is assessed to be worth around $48.6 billion by 2025, which was assessed to be $34 billion in 2018.

9.2.2   WHY MEDICAL IMAGING SO IMPORTANT?

Medical imaging used for diagnosing is considered an essential part of the validation of a diagnosis, and the recording of many ailments and disorders. Fine quality imagery continues to lead in many computer vision techniques. Not only does it facilitate patient decision-making, but it can also minimize unnecessary medical procedures, too; for example, the use of medical imaging technologies, such as ultrasound and MRI, may help prevent surgical procedures.

Earlier diagnosis include experimental methods in detecting an elderly person's problems, children with chronic pain, pre-diabetes, and tumor detection. Medical imaging made easy availability of critical healthcare information to assist in diagnosing diseases such as tumor growth, lung fever, hemorrhaging, traumatic brain damages, and much more.

The National Bureau of Economic Research Study illustrated an increase in individual life probability with the iterative practice of medical imaging (Sorenson et al., 2013). Therefore, reasonable assumptions can be made that medical imaging diagnosis and remedies will prevent aggressive and severe procedures. The issues posed by these measures should be minimized to reduce the invited costs and the time taken by these techniques. A Harvard scholar reported in his study that $385 spent on these techniques can save nearly $3000 for a one-day hospital stay.

A study by Linder and Schiska (2015) announced that diagnostics through clinical imaging assisted clinical experts with recommending medications and henceforth breast cancer mortality has diminished by 22% to 34%, with the appearance of these procedures. In addition, the early therapy in halting blood clusters reduced a 20% mortality rate brought about by colon cancer (Rawla et al., 2019). Early detection through useful imaging techniques, therefore, motivated the specialists to early analyze illnesses and the patients by the opportunity to fight for a more extended lifetime. Being an erratic technology, clinical imagery is developing gradually, with the headway in Computer Vision. Thus, in the coming years, its utilization will keep on developing with the expanding quantities of Computer Vision researchers and well-being experts simultaneously to propel clinical imaging.

9.2.3   Who Does It and for Whom?

Medical practitioners use diagnostic imaging to assess rehabilitation procedures depending on the organ state. The choice of imagery is made based on the body being studied, and the patient's health problems. Therefore, patients are checked beforehand whether their body responds adversely to the rays emitted by imaging devices to ensure that the least amount of emission of rays available is used for the procedure. Besides, adequate care is taken to prevent effects on remaining portions of the body.

Physicians, Patients, and Computer Vision scholars are the consumers of the medical imaging described as follows:

Medical professionals such as physicians and clinicians operate it for structure review and prescribe the essential medication and also maintain a graphical database for upcoming references in another clinical occurrence as well.

Patients are the sufferers seeking services from the final report produced from the medical images.

Computer Vision scholars using conventional image processing or new, Deep Learning-based methods can label the image data set from underlying patterns with the aid of medical specialists.

9.2.4   HOW IT IS DONE?

Medical imaging is diagnostic imaging that includes radiology. The following radiological techniques are included

Radiography
MRI - Magnetic Resonance Imaging
Ultrasound Endoscopy
Thermography
Nuclear Medicine Imaging
Tomography

9.2.5   WHAT IS MEDICAL IMAGE CLASSIFICATION?

Object classification is a challenging one for the machines. Image classification is a complex process that includes various procedures like pre-processing of images, imaging sensors, segmentation of objects, and extraction of features, object detection, and categorization of objects. Predefined patterns maintained in a database are used to compare an object to be classified into the relevant class. Image Classification stays a basic and motivating assignment in different use-cases including distant detecting, vehicle route, biomedical imaging, video-observation, biometry, modern visual examination, robot route (Han & Davis, 2011).

Clinical Image Classification is the way to group clinical images into various classes so that specialists can easily analyze an illness or use it for additional examination. Regardless of the fact that an enormous number of examination papers are distributed in this field, clinical images originated from different sources may vary in an area of focus, pixel intensities, textures, and color balance. If traditional features are only used for classifying medical images, characterizing certain classes efficiently would be tedious (Zare et al., 2013). Deep Learning has been one of the most recent fields of study in computer science and its applications in the last few years. Hinton et al. (2006) were the first to discuss the framework of the Deep Learning model. Several other Deep Learning models for solving image problems have now been proposed.

9.2.6   WHY DEEP LEARNING OVER CONVENTIONAL METHODS

The healthcare business is the main area where clinical specialists make most of the understandings of clinical data. Only a few experts are quite aware of the interpretation of medical images because of its complexity, different kinds of parameters, and most significant fundamental information on the field.

With the advent of medical image acquisition devices as discussed earlier, challenges in collecting data have been reduced over time. Thus, we are in a period of rapid growth in the creation of medical imaging, as well as demanding and insightful analyses about it. Moving to Big Data increases the burden of medical experts for analyzing that data to get insights out of it and may likely be prone to human error.

Besides, conventional machine learning algorithms are not able to understand the complexity of such healthcare-related problems; they mostly rely on the features extracted by medical experts of the concerned field.

Being a work-intensive process, traditional methods of learning were not accurate, as the patients' data differ and knowledge of data analysis often differs from medical expert experience.

The customary AI calculations such as Logistic Regression, Support Vector Machines, K-Nearest Neighbors, Decision Trees, and others are prepared to learn crude image information without learning obscure examples. Data preprocessing is likewise time consuming as they are centered around the data given by the clinical specialists.

Deep Learning-based algorithms, on the other hand, have made tremendous strides in Computer Vision for identifying unknown patterns and extracting descriptors from raw image data. Effective diagnosis by a clinician has improved in recent times with the help of finer data and the monitoring of extracted features. Common Deep Learning strategies include algorithms such as Convolutional Neural Networks (CNNs), Recurrent Neural Networks (RNNs), Long Short Term Memory Networks (LSTMNs), Generative Adversarial Networks (GANs), etc. These algorithms do not require manual pre-processing of the raw image.

9.3 OVERVIEW OF DEEP LEARNING

9.3.1 FUNDAMENTALS OF DEEP LEARNING

Deep Learning models, which can have numerous regular and non-regular processing units, can learn high-level abstraction of the information organized in a more profound fashion (Deng et al., 2014). It is an improvement over the Artificial Neural Network (ANN) and is comprised of a more hidden layer that takes into consideration higher deliberation levels and better image acknowledgment. This part gives a prologue to the ideas and methods of DL.

9.3.1.1 Aspects of Deep Learning

Deep Learning can be characterized by the following key features

Multiple levels of representations in Hierarchy
Multifunctional neural networks
Exercising massive neural networks
Non-linear multiple transformations
Pattern Identification
Extraction function
High-level model for data abstraction

9.3.1.2 Drivers of Deep Learning

The drivers of Deep Learning that have empowered specialists to completely use the guaranteed capability of Deep NNs can be listed as follows:

Multi-layered learning network available
Big Data Management Potential
Expanded use of high-performance graphics processing units (GPU)
An enhanced data scale and neural network size
Improved Neural Networking performance
A large amount of labeled data is available
GPUs' capacity to conduct parallel computing

9.3.2 DEEP LEARNING ARCHITECTURES

Artificial Neural Networks (ANN) copy the human sensory system, genuinely and theoretically. Perceptron is a human brain framework dependent on the most punctual Neural Networks. It has an input layer straightforwardly associated with the output layer, which is excellent in the classification of linearly seperable models. A Neural Network that has a layered draftsman was acquainted with settle more unpredictable examples.

Deep Learning is the enhancement of the ANN and consists of a hidden layer, which allows for higher abstraction levels and better image recognition. DL is one of the most exciting leading technologies in the field of automation. At present, some Computer Vision systems focusing on Deep Learning work much better than humans, including detecting signs of blood cancer in blood samples as well as tumors using MRI scans. Because of its recent and unprecedented outcome, it has become a widely-used method for various functions such as object recognition, speech recognition, face detection, and medical imaging.

In DL, there are a large variety of architectures and algorithms. Here in this section, a few Deep Learning architectures are discussed briefly. LSTM and CNN are the two oldest and the most widely used techniques.

The Deep Learning models—Repetitive Neural Networks (RNN), Long Short-Term memory (LSTM), and Gated Recurrent Units (GRUs), Convolutional Neural Networks (CNN), Deep Belief Networks (DBN), and Deep Stacking Networks (DSN) are the most famous strategies. A portion of the commonplace uses of various Deep Learning models are given in the graph below. See Figure 9.1.

9.3.2.1 Deep Neural Networks (DNN)

A Neural Network, a layered architecture that consists of an input layer, an output layer, and hidden layers was introduced to solve more complex patterns. It has interconnected neurons presented, that accept input and process the input data to some degree, and eventually forwards the current layer output as input to the next layer. Figure 9.2 shown below is a typical neural network architecture.

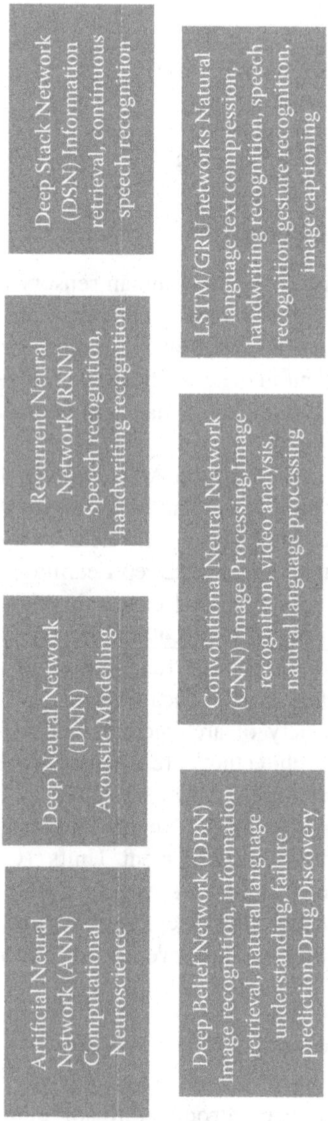

FIGURE 9.1 Use cases of deep learning architectures.

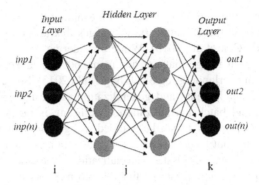

FIGURE 9.2 Neural network architecture.

A neuron sums up the information and applies the initiation capacity to the summed up information, and finally sends the yield to the following layer. The expansion of extra hidden layers allows you to manage a complex non-direct relationship catch as a shrouded layer. This sort of neural engineering is called a Deep Neural Network (DNN).

A DNN hierarchically stacks neurons in a multi-layer fashion. When the number of layers exceeds more than 1,000 gigantic modeling power is made that deep networks can remember all possible mappings from input to output with a necessarily large information database after good training and it can create intelligent predictions such as interpolations and extrapolations for hidden cases. In this way, Deep Learning-based algorithms have a significant effect on Computer Vision and medical imaging. A similar effect occurs in fields such as text recognition, voice recognition, and others.

9.3.2.2 Convolutional Neural Networks

In Convolutional Neural Networks (CNNs), the variety of multi-layer perceptrons is likely to accept patterns directly from raw pixels of visual images. CNN has convolutional layers, pooling layers, completely associated layers, and a softmax layer. Consolidating convolutional layers and pooling layers will assist us with removing highlights. The softmax layer is seen as a classifier. Every hub, called a neuron in a Deep Neural Network, is coordinated by an enactment work, used to control the yield. There are numerous activation functions utilized in Deep Learning methods such as linear, sigmoid, tanh, and Rectified Linear Units (ReLUs).

Another significant interaction in Convolutional Neural Networks is pooling, which is a non-straight downsampling strategy. The various sorts of pooling layers utilized are stochastic, max, and mean pooling. Max pooling is the way to take local maxima for each block made by isolating the input image into rectangular blocks. Max pooling has two advantages, eliminating minimum values to shorten computation and reducing the dimension of intermediate feature maps, maintaining translational invariance and robustness of the framework. Mean pooling is another type of pooling measure, by which the sub-blocks are made dependent on the mean worth. In stochastic pooling, the initiation work is arbitrarily chosen inside the dynamic actuation pooling limit. Besides downsampling the feature maps, pooling layers are also used in the classification of images that are translational and rotational invariance(LeCun et al., 2015). While considering weak spatial data neighboring the areas of an image that are dominant, pooling layers can also help in learning features on overlapping regions (Ding et al., 2015). Some other techniques used by DL for better learning and generalization of the model are L1, L2 regularizer, dropout, and batch normalization. Overfitting is the main disadvantage while training a DCNN and it can be avoided using a dropout layer (Srivastava et al., 2014). The dropout layer is used for regularization where randomly selected connections are dropped out. Batch normalization is another method used for regularization where the input data is divided into mini-batches. Besides regularization, this technique also improves the training speed (Loffe & Szegedy). As the performance of DL mainly depends on a large data set, several augmented methods are used for cases where limited data is available (Kooi et al., 2017). Augmentation methods may include random cropping, color jittering, image flipping, and random rotation (Perez & Wang). See Figure 9.3.

CNN-based architectures have become more popular for processing digital images, especially in the Computer Vision and medical image analysis fields (Premaladha and Ravichandran, 2016; Kharazmi et al., 2018; Wang et al., 2018). Various types of CNN architectures include AlexNet, LeNet, Faster R-CNN, GoogleNet, ResNet, VGGNet, and ZFnet, etc. LeNet-5 (LeCun et al., 1998) is the first CNN model proposed for recognizing handwritten text.

9.3.2.3   Recurrent Neural Network (RNN)

The RNN is considered as the foundational network architecture from which to build other Deep Learning architectures for modeling sequential data such as time series or natural language. The basic difference of RNN from a multi-layer network is the feedback connections within the network that maintain the memory of previous inputs. RNNs have a type of chain of rehashing modules of Neural Networks, for example, a solitary tanh layer. Unfortunately, RNN cannot learn long-haul reliance issues. See Figure 9.4.

Long Short-Term Memory and Gated Recurrent Unit are two of the standard varieties of RNN for determining the vanishing gradient issue produced by lengthy input sequences.

Long Short Term Memory Deep Networks (LSTMs) – an exceptional sort of RNN – were presented by Hochreiter and Schmidhuber (1997). LSTMs are unequivocally intended to evade the drawn-out reliance issue and vanishing gradient problem encountered while training a standard RNN. LSTM is made of repeating modules with different structures. This chain-like structure has four interaction layers, instead of using a single neural network layer.

In Figure 9.5, every line transfers vector value, from the output of one neuron to the input of other neurons. The gray circle is a neural network layer that performs the pointwise operations as vector addition. Merging the lines indicate concatenation and line forking denotes copied content being shared to different locations.

Gated Recurrent Units (GRUs) are an equally effective variation of LSTMs (Cho et al., 2014). They have minimal gating units with fewer parameters,

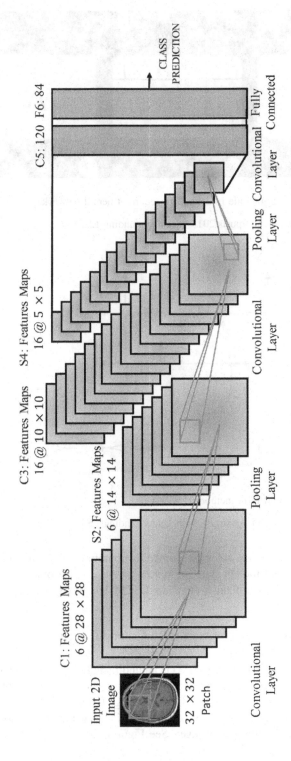

FIGURE 9.3 A typical convolutional neural network (LeNet-5) architecture for medical image classification.

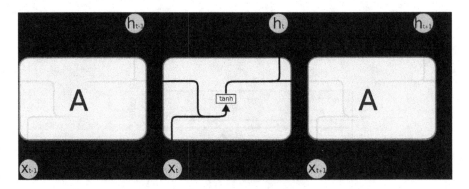

FIGURE 9.4   Repeating module in a standard recurrent neural network.

(Source: http://colah.github.io/posts/2015-08-Understanding-LSTMs/).

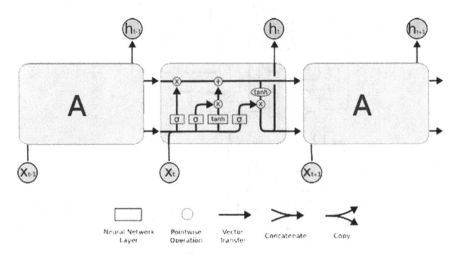

FIGURE 9.5   Long short term memory (LSTM).

(Source: http://colah.github.io/posts/2015-08-Understanding-LSTMs/).

compared to LSTMs. Update gates are formed by combining forget and input gates. Also, the cell state and hidden state are united to make some other changes. The resultant model is simpler than a standard LSTM but has gained popularity in recent years.

9.3.2.4   Deep Boltzmann Network (Also Called Restricted Boltzman Machine)

A Restricted Boltzmann Machine (RBM) is an undirected graphical model valuable for dimensionality decrease, characterization, relapse, community-oriented separating, highlight learning, and theme demonstrating. RBN is a shallow two-layer network, of which the principal layer of the RBM is known as the obvious or info layer and the second is the hidden layer. See Figure 9.6.

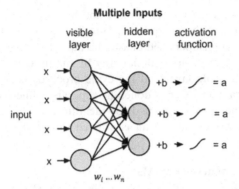

FIGURE 9.6   Restricted boltzman machine architecture.

RBMs share a similar idea of an autoencoder, but one aspect that differentiates RBMs from other autoencoders is that it has two biases. One drawback of an RBM is that there is no intralayer communication. Each neuron makes hypothetical decisions to decide whether to transfer input data or not. See Figure 9.7.

9.3.2.5   Deep Belief Networks (DBNs)

This multilayer network configuration ordinarily contains an imaginative preparing calculation. RBN is the structure square of Deep Belief Networks. The hidden layers in each subnetwork fill in as an information layer for the following layer. The greedy methodology utilized in each layer maximize straightforwardly the likelihood. The initial process makes the trainig cycle computationally costly.

In a DBN, the info layer has crude information sources and unique portrayals of these information sources are learned by the hidden layers. The yield layer

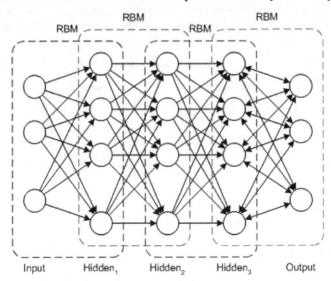

FIGURE 9.7   Deep belief network (DBN).

actualizing the network characterization is not the same as different layers. Two phases of preparing a DBN incorporate pre-training phase and fine-tuning phase.

In Pre-training, an unsupervised phase, the input is reproduced by each RBM unit. The RBM is engaged in a similar way by the initially hidden layer as the information or noticeable layer and the cycle proceeds until each layer is prepared. When the training phase is done, the supervised fine-tuning phase begins, where the output neurons are named for identification. Either the back propagation or gradient descent algorithm is utilized to update the data to complete the training process.

9.3.2.6   Deep AutoEncoder (DAE)

A Deep Autoencoder is an unsupervised Deep Neural Network that recreates the contribution as precisely as can be expected. Autoencoders are typically used as dimensionality reduction network with a small amount of loss of data. Autoencoders contain an encoder and a decoder. The encoder packs the information and the decoder decompresses the information to make the remaking of the contribution as close as could be expected. See Figure 9.8.

Deep autoencoders learn non-linear transformations with activation functions and also multiple layers. Instead of learning dense layers, convolutional layers can be used for learning as it is good for video, series data, and images. Rather than learning huge transformations with Principle Component Analysis (PCA), a learning autoencoder is efficient. Autoencoders can achieve transfer learning by using pre-trained layers from another model for enhancing encoders or decoders

9.4   DEEP LEARNING FOR MEDICAL IMAGE CLASSIFICATION [LITERATURE REVIEW]

The vital point of Image Classification is to recognize peculiarities. The significance of classifying clinical information is to possibly assemble a prescient model or

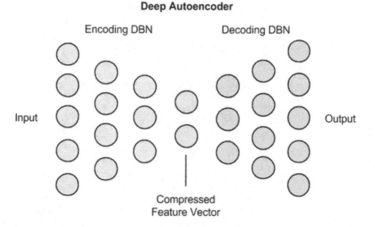

FIGURE 9.8   Deep auto encoder (DAE).

framework to diagnose the type of sickness from typical subjects or to assess the phase of the illness. Deep Learning gives a significant way to deal with the latest analytic techniques. Deep Learning frameworks in the medical care sector is the key to diagnos disease and recommend customized treatment. Deep Learning innovation was first utilized in the field of Medical Image Clasification. Different radiological imaging methods have produced a good amount of data, but we still lack important user data that can be integrated into the Deep Learning system. Let us address the breakthroughs in medical image classification accomplished using Deep Learning.

9.4.1 DEEP LEARNING FOR DIABETIC RETINOPATHY

Diabetes Mellitus, a metabolic problem has two sorts of issues, Type-1 and Type-2 diabetes. The latter being when the pancreas cannot create sufficient insulin, and the former is when the body does not respond to insulin, both types lead to high levels of glucose. Diabetic Retinopathy is an eye issue brought about by diabetes, which prompts lasting visual deficiency with the seriousness of the phase of diabetes.

According to a World Health Organization (WHO) study, the quantity of diabetic patients worldwide has expanded from 108 million to 422 million people within the last 35 years. This metabolic disorder is increasing in low- and medium-livelihood nations.

Impaired vision is mainly caused by diabetic retinopathy, which is a result of retinal blood vessel damage. Diabetes is the main cause of around 2.6% of global blindness.

The seriousness of diabetic retinopathy can be limited and it very well may be cured if detected early by a retinal screening test. Identifying diabetic retinopathy physically is tedious because of the inaccessibility of machines and the abilities required for the test. The early side effects of this issue can not be perceived, for the most part. This creates an issue for ophthalmologists to investigate the fundus images, which is tedious and delays satisfactory and ideal treatment for people with this illness. Deep Learning strategies have given confident outcomes for the robotized diagnosis of diabetic retinopathy.

Gulshan et al. (2016) proposed a Deep Convolutional Neural Network (DCNN) for computerized location of referable diabetic retinopathy from retinal fundus photos. EyePACS-1 dataset –with 9,963 images from 4,997 patients – and the Messidor-2 dataset – with 1,748 images from 874 patients – are the two enormous datasets used to prepare this Deep Neural Network. This paper has achieved 93.4% particularity and 97.5% affectability on the EyePACS-1 dataset, and 93.9% explicitness and 96.1% affectability on the Messidor-1 dataset. Initiation v3 architecture (Szegedy et al., 2016) was the particular Deep convolutional Neural Network used in this study. See Figure 9.9.

Operating closely with both Indian and U.S. doctors, the Google AI team created a dataset of 128,000 images with several members from a panel of 54 ophthalmologists, who evaluated each image. A specific type of DNN was trained to categorize referable diabetic retinopathy. The performance of the algorithm was

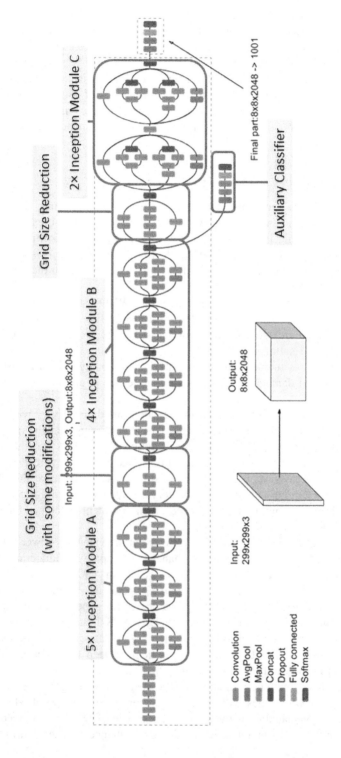

FIGURE 9.9 The inception-v3 architecture proposed by Szegedy et al.

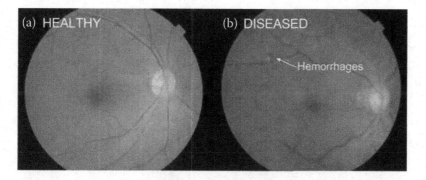

FIGURE 9.10 Retinal fundus photographs to screen DR. Healthy Retina (A) diseased retina with RDR (B) red spots are present due to several hemorrhages in B.

evaluated on two individual clinical validation sets (totaling 12,000 images approx.,), decided by the U.S. board-certified ophthalmologists who are more consistent among the 54 doctors. This team (Kathirvel, 2016) achieved a mind-blowing performance yielding an F1-score of 0.95 comparatively higher than the median F1-score of 0.91 achieved by eight ophthalmologists. See Figure 9.10.

Peng et al (2016) developed a Deep Learning model based on a DNN using public datasets such as Google Kaggle, DRIVE, and STARE for the lesion-based classification of fundus images into five categories including, mild, moderate, severe, and proliferative DR. The network design using dropout techniques achieved a significant accuracy of 94%. See Figure 9.11.

Recently, gold standard automatic DR diagnosis methods have been developed. This has helped doctors to diagnose more patients and increased the diagnosis process that resulted in minimizing the treatment lagging period. Collaborating with clinicians and scholars, Google is working intensely to update the examing process worldwide, with the expectation that both patients and doctors can profit maximally from these approaches.

9.4.2 DEEP LEARNING FOR THE DETECTION OF HISTOLOGICAL AND MICROSCOPIAL ELEMENTS

Histological components research is the investigation of analyzing cells and the gathering of cells and tissues under a magnifying lens. Microscopic imaging and stains are the strategies used to break down the cell and tissue-level microscopic variations. This procedure includes steps like tissue obsession, example move to tapes, tissue preparing, segment, staining, and optical microscopy.

Skin malignant growths such as squamous cell carcinoma and melanoma, gastric disease or gastric carcinoma, middle gastric sores also called precancerous gastric epithelial metaplasia, breast cancer, irresistible illnesses like jungle fever and intestinal parasites, and many more tissue-related infections can be recognized by this imaging strategy. Microscopic imaging of blood smear tests is the standard technique for diagnosing intestinal illnesses brought about by Genus

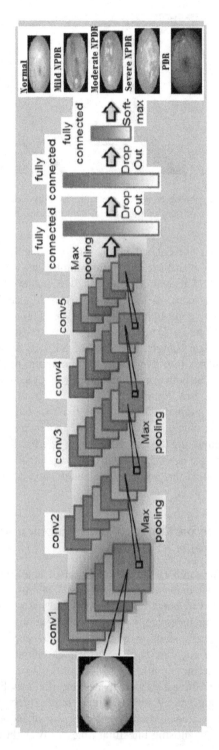

FIGURE 9.11   DCNN architecture used by Kathirvel et al.

Plasmodium parasites. Smear test microscopy and fluorescent auramine-rhodamin stain or Ziehl-Neelsen stain are the standard strategies for distinguishing mycobacteria in sputum tests, which is the primary driver for tuberculosis.

CRCHistoPhenotypes-Labeled Cell Nuclei Data was a datset issued by the Department of Computer Science from Warwick University in 2016. One hundred colorectal adenocarcinomas stained images in histology were included in the dataset, in which a total of 29,756 labeled cell nuclei images were used for detection purposes, and out of them, 24,444 nuclei were identified with labels, namely epithelial nuclei, fibroblast nuclei, inflammatory nuclei, and miscellaneous nuclei.

Sirinukunwattana et al. (2016) proposed a spatially-constrained CNN for detection and a CNN with softmax and ensemble model for classification of colon cancer using the CRC HistoPhenotypes datasets. Combining these two models, research based on tissue structure and its constituents will ultimately come to be a valuable method to understand the context of the cyst in a better way. See Figure 9.12.

The paper (Bayramoglu & Heikkilä, 2016) claimed to achieve the major bottleneck faced in training a deep convolutional neural network model provided with limited availability of training data. The author proposed transfer learning to overcome this issue of limited data for the classification of cell nuclei using histopathology images. The author compared the performances of transfer learning and learning from scratch. Still, the author obtained better performance and evaluated different deep neural networks that are trained using natural and face images.

Quinn et al. (2016) assessed the presentation of Deep Convolutional Neural Networks on three distinctive infinitesimal imaging errands: analysis of jungle fever in thick blood spreads, tuberculosis in sputum tests, and intestinal parasite eggs in feces tests. This paper discussed plasmodium in thick blood smear images; tuberculosis bacilli in sputum tests; and eggs of hookworm, Taenia, and Hymenolepsis

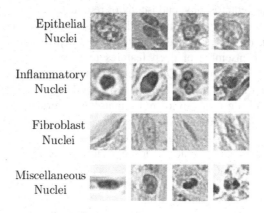

FIGURE 9.12 CRCHistoPhenotypes – labeled cell nuclei dataset (Sirinukunwattana et al., 2016).

nana in feces tests. Robotized microscopic investigation has been performed by recognizing bounding boxes around each object of revenue in each image and accomplished a high AUC of 1.00 for plasmodium images to distinguish malaria, 0.99 to identify tuberculosis, and 0.99 for intestinal parasites identification. The four hidden layers of DCNN utilized in this work are Convolution layer: 7 filters of size 3 × 3. See Figure 9.13.

> Pooling Layer: max-pooling, factor 2.
> Convolution Layer: 12 filters of size 2 × 2.
> Fully Connected Layer, with 500 hidden units.

Malaria diagnosis is extremely important and necessary. A study by the WHO in 2018 estimates that around 228 million malaria cases were registered globally, out of which a count of 405,000 deaths occurred due to malaria. Children under 5 years old are the ones most affected by this disease and accounted for 67% of total deaths by malaria worldwide in 2018.

9.4.3 DEEP LEARNING FOR GASTROINTESTINAL DISEASE DETECTION

The gastrointestinal tract is responsible for the assimilation of food and nutrients ingestion from mouth to butt. The empty organs included in the tract are mouth, throat, duodenum, small and large intestine, and the solid organs include the liver, pancreas, and gall bladder. The upper GI tract includes the throat, stomach, and duodenum, while the lower GI tract consists of the small intestine (small bowel) and large intestine (colon).

Problems, such as, irritation, hemorrhage, disease, and cancer in the gastro-intestinal tract are the main causes of indigestion and improper absorption. Peptic or stomach ulcer is the bleeding caused in the upper GI tract. Tumors, gastric cancer, and colonic diverticulitis cause large bowel hemorrhage. Because of abnormal blood vessels, celiac sprue, Crohn's illness, gastric tumors, peptic ulcers, and GI bleeding are issues that trouble the small intestine. Current imaging innovations such as endoscopy, enteroscopy, endoscopy of remote cases, tomography, and MRI play a huge role in the determination of these gastrointestinal tract issues.

Gamage et al. (2019) proposed an automated Deep Ensemble Network archi-tecture trained on ensemble of deep features to classify GI tract diseases. This model combined pre-prepared DenseNet-201, ResNet-18, and VGG-16 CNN models as the feature extractors followed by a Global Average pooling (GAP) layer with a promising accuracy of more than 97%. Further, the processing time is reduced using a dimensionality reduction technique (SVD-Singular Value Decomposition) without affecting classification accuracy. The KVASIR dataset used in this research incorporates endoscopic images of the inward GI tract that were categorized into eight unique classes as three essential anatomical tourist spots, three significant clinical discoveries, and two classifications of endoscopic images identified with polyp expulsion.

Petscharnig et al. (2017) prepared an Inception-like CNN design and a fixed-crop information expansion plot for testing gastrointestinal infection and anatomical

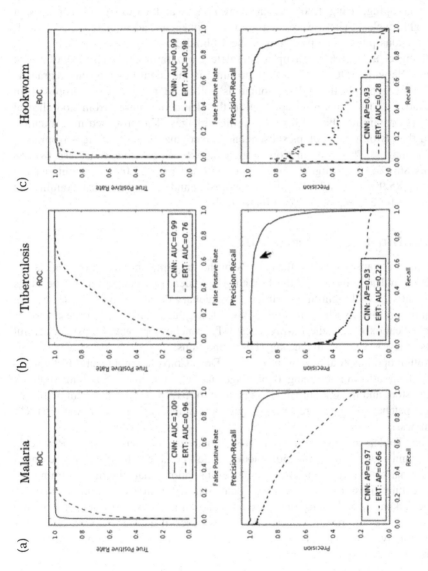

FIGURE 9.13 ROC and precision-recall for detection of malaria, tuberculosis, and intestinal parasites.

milestone arrangement for the Medico task at Medieval, 2017. This engineering depended on Google LeNet and worked to save the number of teachable boundaries and keep their overhead figuring costs small.

Komeda et al (Komeda et al., 2017) proposed a CNN-CAD framework utilizing routine colonoscopy-based engineering for the fast determination of colorectal polyp grouping, using 1,200 colonoscopy images and 10 extra video images of untaught cycles. They were analyzed as either an adenomatous or a non-adenomatous polyp. The precision of the 10-hold cross-approval was 0.751.

Zhang et al. (2016) built up a completely programmed Deep Convolutional Neural Network (DCNN)-based algorithm for the identification and character-ization of hyperplastic and adenomatous colorectal polyps. They proposed an innovative transfer learning application that learned features from non-clinical datasets containing about 1.4 to 2.5 million images. The proposed model antici-pated the histology of polyps subsequent to recognizing polyp images from non-polyp images. This examination brought about comparative exactness (87.3% versus 86.4%) yet, a higher review rate (87.6% versus 77.0%) and a higher ex-actness (85.9% versus 74.3%), when contrasted and when the visual examination was done by endoscopists. See Figure 9.14.

9.4.4 DEEP LEARNING FOR LUNG DISEASE

Asthma, a collapse of part or the entirety of the lung (pneumothorax or atelectasis), swelling and aggravation in the bronchial cylinders that convey air to the lungs (bronchitis), COPD (chronic obstructive pulmonary disease), and Lung cancer are the most well-known lung illnesses. Blood tests, pneumonic capacity tests (spiro-metry), beat oximetry, chest x-ray, chest CT, bronchoscopy with biopsy, or careful biopsy may assist with diagnosing these conditions.

Anthimopoulos et al. (2016) proposed a Deep Convolutional Neural Network to group lung infection utilizing two image datasets, the interstitial lung illnesses (ILDs) store and the CT outputs of an element of 512×512 each. A sum of 14,696 image patches used to prepare the network were removed from the first CT examines.

Van Tulder and de Bruijne (2016) proposed the convolutional Restricted Boltzmann machine classification model, which combines both generative and a discriminative learning objective giving examinations that highlight learning for lung texture classification and airway detection in CT images. From the first CT images taken, 32×32 image patches were chosen along with a framework with a 16-voxel cover to prepare thenetwork. In this paper, it has been demonstrated that discriminative learning can assist an unsupervised learner in learning optimized filters for classification.

Bharati et al. (2020) proposed a novel cross breed Deep Learning structure, VDSNet for the grouping of lung sicknesses utilizing the Kaggle-NIH chest x-ray image dataset. VDSNet indicated the best precision of 73% while approving the whole dataset, similarly higher than vanilla gray (67.8%), vanilla RGB (69%), half breed CNN VGG (69.5%), essential CapsNet (60.5%), and adjusted CapsNet (63.8%). Further, while considering the example dataset, VDSNet accomplished an

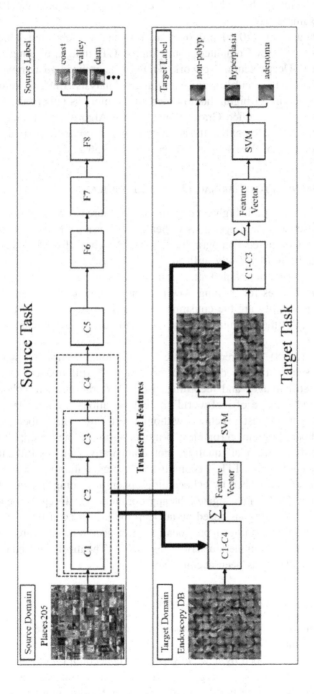

FIGURE 9.14 The pipeline of the method proposed by (Zhang et al., 2016). endoscopy DB-endoscopic images collected. C-Convolutional layer and F-Fully connected layer.

approval precision of 70.8%. Then again, 431s is the preparation time needed by this structure for handling the full dataset, which is a lot higher than the 19s time needed for the example dataset.

Lakshmanaprabu et al. (2019) proposed an imaginative robotized conclusion arrangement framework for Computed Tomography (CT) images of lungs. In this paper, the Optimal Deep Neural Network (ODNN) is utilized to remove Deep highlights from CT images and Linear Discriminate Analysis (LDA) is utilized to lessen the dimensionality of highlights to order lung tumors as either malignant or benign. Additionally, a Modified Gravitational Search Algorithm (MGSA) is utilized for improvement to accomplish the accompanying outcomes: a particularity of 94.2%, the affectability of 96.2%, and a precision of 94.56%.

9.4.5 Deep Learning for Cardiac Disease Classification

Cardiovascular illnesses rank high among the main sources of mortality on the planet. Vascular, ischemic, and hypertensive heart illnesses are a portion of the sorts of heart infections. Computed Tomography (CT) and Magnetic Resonance Imaging (MRI) checks are the most broadly utilized for cardiovascular imaging. Labor extensive for manual annotation, failed attempt in differentiating views at once, usage of only exemplary images for training, low precision, specific client-based equipments, and incompatability are the constraints of utilizing AI strategies for multiview classification. All these limitations make it a tedious task for epidemiological examinations.

Madani et al. (2018) prepared a unique convolutional Neural Network that characterizes 15 standard perspectives (12 recordings, 3 still images) at a time, based on named still images and recordings from 267 transthoracic echocardiograms that captured a range of real-world clinical variations. This model accomplished 97.8% in general test exactness without overfitting. This paper classifies each kind of echocardiogram chronicles: b-mode, m-mode, and Doppler; static images and recordings from all qualities got comparative with a full standard transthoracic echocardiogram-TTE, accomplished a bigger number of precisions than the exactness accomplished by echocardiographers for doing likewise task.

Amiriparian et al. (2018) investigated the appropriateness of Deep Autoencoders that learn feature representations based on auDeep toolkit for acoustic information classification. The study cataloged phonocardiogram accounts into three distinct classes, ordinary, mild, and moderate/serious anomalies. The most elevated recall of 47.9% on the test dataset has been accomplished.

9.4.6 Deep Learning for Tumor Detection

A tumor neoplasm is the unusual development of cells of any piece of the body shaping a different mass of tissue. Body cells go through a pattern of developing, maturing, blurring, and subbed by new cells finally. On account of a tumor or some other type of malignancy, this cycle gets interfered. Benign (non-destructive) and Malignant (dangerous) are the two kinds of tumors. A benign tumor is not much serious and sticks to one section without spreading to other body parts. On the other hand, a malignant tumor is dangerous and spreads to residual pieces of the body, making both diagnosis and treatment troublesome.

Shen et al. (2019) prepared a Deep Convolutional Neural Network on mammographic images for improved recognition and grouping of breast cancer. Digital Database for Screening Mammography (DDSM) is an old-fashioned version containing the lossless-JPEG arrangement of digitized film mammograms. So a refreshed adaptation, CBIS-DDSM having standard DICOM configuration of mammographic images were utilized. The dataset utilized included 2,478 mammographies screened in both craniocaudal view (CC) and mediolateral diagonal view (MLO) from 1,249 patients. In this examination, each view is an image. The study prepared the network in two stages, in which the initial stage was to distinguish Region of Interest (ROI) utilizing a patch classifier and the subsequent advance used Resnet-50 and VGG-16 classifier to group the identified ROI images into various classes, namely, background, malignant mass, benign mass, malignant calcification, and benign calcification. The training data and test data were randomly created by spliting the dataset in the proportion of 85:15. The validation datais made by further splitting the training data in the proportion of 90:10. In view of the proportion proposed the complete number of images in the training set, validation set, and test set are 1,903, 199, and 376, individually. See Figure 9.15.

The engineering utilized by (Shen et al., 2019) changing a fixed classifier over to a start to finish teachable entire image classifier utilizing an all convolutional plan. The capacity f was first prepared on patches and afterward refined on entire images. See Figure 9.16.

9.4.7 DEEP LEARNING FOR ALZHEIMER'S AND PARKINSON'S DETECTION

Alzheimer's disease (AD), a typical type of feeble dementia is a reformist degenerative cerebrum issue prompting perpetual cognitive decline and upsetting the capacity to complete straightforward assignments. Precise conclusion of AD and

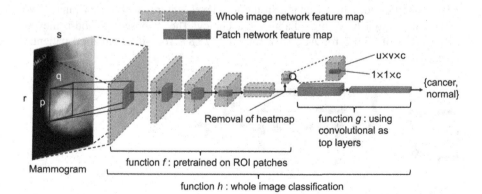

FIGURE 9.15 DCNN architecture (Shen et al., 2019) for the detection and classification of breast cancer.

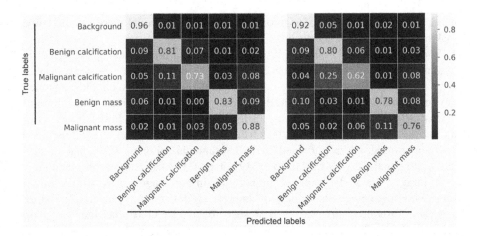

FIGURE 9.16   Confusion matrix analysis (Shen et al., 2019).

prodromal stage like Mild Cognitive Impairment (MCI) assume a critical part in patient consideration chiefly at the beginning phase of the sickness.

Parkinson's is, likewise, a central nervous system disorder influencing basal ganglia in the cerebrum. It causes a progressing decline in the motor system causing indications such as quakes in the hand followed by lethargic body movement, muscle firmness, and poor balance.

Sarraf & Tofighi (2016) proposed CNN and LeNet-5 architecture to classify fMRI data of Alzheimer's subjects from normal controls. This paper suggested that CNN extracted the shift and scale-invariant features and classified fMRI of Alzheimer's patients from healthy controls. They suggested this as the most powerful method to differentiate diseased data from normal controls. This experiment resulted in a mean accuracy of 96.85% using 270,900 images for training and 90,300 images for testing. See Figure 9.17.

Suk & Shen (2013) used a Stacked Autoencoder for AD/MCI classification using the ADNI dataset. They believed that a robust model can be built by combining the latent complicated patterns inbuilt in the low-level features with the original low-level features to achieve high diagnostic accuracy. The proposed method achieved an accuracy of 95.9% for AD, 85% for MCI, 75.8% for MCI-converted patients, respectively.

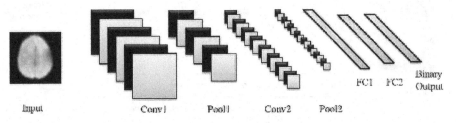

FIGURE 9.17   LeNet-5 architecture (Sarraf and Tofighi, 2016) for the classification of alzheimer's disease.

Sivaranjini & Sujatha (2020) proposed a Deep Learning model that ordered the MR images of sound subjects and Parkinson's illness subjects. This model utilized the Convolutional Neural Network design AlexNet to analyze PD and accomplished a precision of 88.9%. Deep Learning models can help specialists in disease identification and produce an unbiased and better grouping of patients.

9.5 CURRENT PROGRESS AND LIMITATIONS OF DEEP LEARNING

Implementing Deep Learning strategies in clinical imaging to diagnose and cure infections is upsetting more in the region of radiology. Deep Learning has demonstrated extraordinary guarantee in some clinical spaces such as ophthalmologic findings, pathology conclusions, and cancer analysis.

Google Health and the National Health Service in the UK have made an agreement for handling 1,000,000 patients' clinical information. The IBM Watson Company has become a main supplier of clinical imaging equipment and programming after procuring Merge Healthcare. The application of Deep Learning algorithms in medical imaging has led to a great improvement in patient care, but many challenges are slowing the progress. Some of the major potential challenges are summarized below.

9.5.1 LIMITED AVAILABILITY OF DATASETS

The primary quality of Deep Learning is naturally learning portrayals of information. Very capable and enormous amounts of datasets are the most required rule for assessing Deep Learning-based model. The premier test in succeeding with a Deep Learning-based Network for Computer Vision is the restricted dataset accessibility of clinical imaging information and is the greatest test for the accomplishment of Deep Learning in clinical imaging.

To handle this test, all-around experienced clinical specialists are needed for the advancement of a colossal dataset for preparing a Deep Neural Network. Likewise, this might be tedious for these specialists while confronting this test. Moreover, a stable dataset is needed for Deep Learning-based calculations in learning the hidden portrayals of the information accurately.

This examination faces a few difficulties while working on a large-scale dataset. Subsequently, the utilization of small datasets can give great accuracy yet it won't be successful in real applications (Bharati et al., 2020).

9.5.2 PRIVACY AND LEGAL ISSUES

Another challenge faced while evaluating a Deep Learning model is data privacy. Medical data compared to real world images is highly confidential and sensitive. Also, medical data sharing is quite complex and challenging. Issues of data privacy should be viewed both technically and sociologically.

The Health Insurance Portability and Accountability Act of 1996 (HIPAA) of the U.S. is a security law enacted to provide privacy protection for patients' health data from being released without their approval or awareness. Hence, maintaining the secrecy of healthcare data is a big problem for data scientists, as mishandling a patient's data may dispute in patient care.

Undertaking differential privacy approaches may restrict the information provided to the network on a demand requirement basis. Sharing sensitive data with restricted exposure is challenging and ends up with many restrictions. Accessing restricted data because of a limitation decreases the number of valuable information. Aside from that, moving to Big Data is a progressive threat to data security.

9.5.3　DATA AND MODEL STANDARDIZATION

1. **Data Standardization** is the demand for Deep Learning in any research areas, specifically in the medical care domain. The motive behind the variations of image data captured among different hardware causes inconsistency of data. The medical care domain is a combination of medical data from various resources. To improve better performance and accuracy, data standardization is most necessary for Deep Learning-based models. Standardization and sharing of patient care data is done by some organizations including HIPAA, HL7, and HITECH which define the guidelines that have to be maintained. Professional opinions on Electronic Health Records are provided by a main certification authority, Authorized Testing and Certifying Body (ATCB).

2. **Mysterious Black Box Problem** New opportunities have been developed by Deep Learning in medical imaging technology. Deep Learning-based algorithms solve many complex problems that traditional machine learning networks failed to achieve. However, a lack of transparency, called a black box problem, is one of the biggest barriers for the efficiency of Deep Learning in problem domains like trading or medical diagnosis. Neural Networks constructed by math concepts are simple but the model gets complicated and indefinable when weight matrices are calculated for deeper layers.

9.6　CONCLUSION

Recently, Deep Learning has offered many automated daily applications and provided significant improvements to conventional Machine Learning-based algorithms. Deep Learning, which imitates the function of the human brain, has already made strides in fields such as speech recognition and cancer detection. However, Deep Learning applications in medical care are still challenging, as medical data is limited with availability and complexity.

The primary quality of Deep Learning is consequently learning portrayals of information. Well-qualified and large datasets are the most required criteria for evaluating a Deep Learning-based model. The foremost challenge in creating a successful Deep Learning-based Network for Computer Vision is the restricted dataset accessibility of clinical imaging information, and the greatest challenge of the accomplishment of Deep Learning is in clinical imaging.

To handle this challenge, all around experienced clinical specialists are needed for the advancement of an enormous dataset for preparing a Deep Neural Network.

Additionally, this might be tedious for these specialists while confronting this test. Moreover, a stable dataset is needed for Deep Learning-based calculations in learning the hidden portrayals of the information effectively.

This research field faces few difficulties while working on the large-scale dataset. Consequently, the utilization of small datasets can give high accuracy; however, it won't be successful in real applications.

REFERENCES

Amiriparian, S., Schmitt, M., Cummins, N., Qian, K., Dong, F., & Schuller, B. (2018, July). Deep unsupervised representation learning for abnormal heart sound classification. In *2018 40th Annual International Conference of the IEEE Engineering in Medicine and Biology Society (EMBC)* (pp. 4776–4779).

Anthimopoulos, M., Christodoulidis, S., Ebner, L., Christe, A., & Mougiakakou, S. (2016). Lung pattern classification for interstitial lung diseases using a deep convolutional neural network. *IEEE Transactions on Medical Imaging, 35*(5), 1207–1216.

Bayramoglu, N., & Heikkilä, J. (2016, October). Transfer learning for cell nuclei classification in histopathology images. In *European Conference on Computer Vision* (pp. 532–539), Amsterdam, The Netherlands.

Bharati, S., Podder, P., & Mondal, M. R. H. (2020). Hybrid deep learning for detecting lung diseases from X-ray images. *Informatics in Medicine Unlocked, 20,* 100391.

Borra , S., Dey, N., & Ashour, A. S. (Eds.) (2017). Classification in BioApps: Automation of decision making. Springer.

Chandrakumar, T., & Kathirvel, R. (2016). Classifying diabetic retinopathy using deep learning architecture. *International Journal of Engineering Research & Technology, 5*(6), 19–24.

Cho, K., Van Merriënboer, B., Bahdanau, D., & Bengio, Y. (2014). On the properties of neural machine translation: Encoder-decoder approaches. *arXiv preprint arXiv:1409.1259.*

Deng, L., & Yu, D. (2014). Deep learning: methods and applications. *Foundations and Trends in Signal Processing, 7*(3-4), 197–387.

Ding, S., Lin, L., Wang, G., & Chao, H. (2015). Deep feature learning with relative distance comparison for person re-identification. *Pattern Recognition, 48*(10), 2993–3003.

Gamage, C., Wijesinghe, I., Chitraranjan, C., & Perera, I. (2019, July). GI-Net: anomalies classification in gastrointestinal tract through endoscopic imagery with deep learning. In *2019 Moratuwa Engineering Research Conference (MERCon)* (pp. 66–71), Moratuwa, Sri Lanka.

Gulshan, V., Peng, L., Coram, M., Stumpe, M. C., Wu, D., Narayanaswamy, A., ... Kim, R. (2016). Development and validation of a deep learning algorithm for detection of diabetic retinopathy in retinal fundus photographs. *JAMA, 316*(22), 2402–2410.

Han, B., & Davis, L. S. (2011). Density-based multifeature background subtraction with support vector machine. *IEEE Transactions on Pattern Analysis and Machine Intelligence, 34*(5), 1017–1023.

Hinton, G. E., Osindero, S., & Teh, Y.-W. (2006). A fast learning algorithm for deep belief nets. *Neural Computation, 18*(7), 1527–1554.

Ioffe, S., & Szegedy, C. (2015, June). Batch normalization: Accelerating deep network training by reducing internal covariate shift. In *International Conference on Machine Learning (PMLR)* (pp. 448–456).

Kharazmi, P., Zheng, J., Lui, H., Wang, Z. J., & Lee, T. K. (2018). A computer-aided decision support system for detection and localization of cutaneous vasculature in dermoscopy images via deep feature learning. *Journal of Medical Systems, 42*(2), 33.

Komeda, Y., Handa, H., Watanabe, T., Nomura, T., Kitahashi, M., Sakurai, T., & Takenaka, M. (2017). Computer-aided diagnosis based on convolutional neural network system for colorectal polyp classification: Preliminary experience. *Oncology*, *93*(Suppl. 1), 30–34.

Kooi, T., Litjens, G., van Ginneken, B., Gubern-Mérida, A., Sánchez, C. I., Mann, R., den Heeten, A., & Karssemeijer, N. (2017). Large scale deep learning for computer aided detection of mammographic lesions. *Medical Image Analysis*, *35*, 303–312. 10.1016/j.media.2016.07.007.

Lakshmanaprabu, S. K., Mohanty, S. N., Shankar, K., Arunkumar, N., & Ramirez, G. (2019). Optimal deep learning model for classification of lung cancer on CT images. *Future Generation Computer Systems*, *92*, 374–382.

LeCun, Y., Bengio, Y., & Hinton, G. (2015). Deep learning. *Nature, 521*(7553), 436.

LeCun, Y., Bottou, L., Bengio, Y., & Haffner, P. (1998). Gradientbased learning applied to document recognition. *Proceeding of IEEE*, *86*(11), 2278–2324.

Linder, J. M. B., & Schiska, A. D. (2015). Progress in diagnosis of breast cancer: Advances in radiology technology. *Asia-Pacific Journal of Oncology Nursing*, *2*(3), 186.

Madani, A., Arnaout, R., Mofrad, M., & Arnaout, R. (2018). Fast and accurate view classification of echocardiograms using deep learning. *NPJ Digital Medicine*, *1*(1), 1–8.

Peng, L., & Gulshan, V. (2016). Deep learning for detection of diabetic eye disease. *Google Research Blog*. https://ai.googleblog.com/2016/11/deep-learning-for-detection-of-diabetic.html

Perez, L., & Wang, J. (2017). The effectiveness of data augmentation in image classification using deep learning. *arXiv preprint arXiv:1712.04621.*

Petscharnig, S., Schoffmann, K., & Lux, M. (2017). An Inception-like CNN Architecture for GI Disease and Anatomical Landmark Classification. In *MediaEval*.

Premaladha, J., & Ravichandran, K. (2016). Novel approaches for diagnosing melanoma skin lesions through supervised and deep learning algorithms. *Journal of Medical Systems* *40*(4), 96.

Quinn, J. A., Nakasi, R., Mugagga, P. K., Byanyima, P., Lubega, W., & Andama, A. (2016, December). Deep convolutional neural networks for microscopy-based point of care diagnostics. In *Machine Learning for Healthcare Conference (PMLR)* (pp. 271–281).

Rawla, P., Sunkara, T., & Barsouk, A. (2019). Epidemiology of colorectal cancer: Incidence, mortality, survival, and risk factors. *Przegląd Gastroenterologiczny*, *14*(2), 89.

Razzak, M. I., Naz, S., & Zaib, A. (2018). Deep learning for medical image processing: Overview, challenges and the future. In *Classification in BioApps* (pp. 323–350).

Sarraf, S., & Tofighi, G. (2016). Classification of Alzheimer's disease using FMRI data and deep learning convolutional neural networks, *arXiv preprint arXiv:1603.08631.*

Shen, L., Margolies, L. R., Rothstein, J. H., Fluder, E., McBride, R., & Sieh, W. (2019). Deep learning to improve breast cancer detection on screening mammography. *Scientific Reports*, *9*(1), 1–12.

Simie, E., & Kaur, M. (2019). Lung cancer detection using convolutional neural network (CNN). *International Journal of Advance Research, Ideas and Innovations in Technology*, *5*, 284–292.

Sirinukunwattana, K., Raza, S. E. A., Tsang, Y. W., Snead, D. R., Cree, I. A., & Rajpoot, N. M. (2016). Locality sensitive deep learning for detection and classification of nuclei in routine colon cancer histology images. *IEEE Transactions on Medical Imaging*, *35*(5), 1196–1206.

Sivaranjini, S., & Sujatha, C. M. (2020). Deep learning based diagnosis of Parkinson's disease using convolutional neural network, *Multimedia Tools and Applications*. *79*(21), 15467–15479.

Sorenson, C., Drummond, M., & Khan, B. B. (2013). Medical technology as a key driver of rising health expenditure: Disentangling the relationship. *ClinicoEconomics and Outcomes Research: CEOR, 5,* 223.

Srivastava, N., Hinton, G., Krizhevsky, A., Sutskever, I., & Salakhutdinov, R. (2014). Dropout: A simple way to prevent neural networks from overfitting. *The Journal of Machine Learning Research, 15*(1), 1929–1958.

Suk, H. I., & Shen, D. (2013, September). Deep learning-based feature representation for AD/MCI classification. In *International Conference on Medical Image Computing and Computer-Assisted Intervention* (pp. 583–590), Berlin, Heidelberg.

Szegedy, C., Vanhoucke, V., Ioffe, S., Shlens, J., & Wojna, Z. (2016). Rethinking the inception architecture for computer vision. In *Proceedings of the IEEE Conference on Computer Vision and Pattern Recognition* (pp. 2818–2826).

van Tulder, G., & de Bruijne, M. (2016). Combining generative and discriminative representation learning for lung CT analysis with convolutional restricted Boltzmann machines. *IEEE Transactions on Medical Imaging, 35*(5), 1262–1272.

Wang, S.-H., Phillips, P., Sui, Y., Liu, B., Yang, M., & Cheng, H. (2018). Classification of Alzheimer's disease based on eight-layer convolutional neural network with leaky rectified linear unit and max pooling. *Journal of Medical Systems, 42*(5), 85.

Zare, M. R., Mueen, A., Awedh, M., & Seng, W. C. (2013). Automatic classification of medical x-ray images: Hybrid generative discriminative approach. *IET Image Processing, 7*(5), 523–532.

Zhang, R., Zheng, Y., Mak, T. W. C., Yu, R., Wong, S. H., Lau, J. Y., & Poon, C. C. (2016). Automatic detection and classification of colorectal polyps by transferring low-level CNN features from nonmedical domain. *IEEE Journal of Biomedical and Health Informatics, 21*(1), 41–47.

10 A Comparative Review of the Role of Deep Learning in Medical Image Processing

Erapaneni Gayatri and S. L. Aarthy

CONTENTS

10.1 INTRODUCTION

Nowadays, in the medical sector, Deep Learning methods act as a dynamic character in the analysis of medical images and identification of diseases. The ease of understanding the levels of images also improved, due to the rapid growth of computer-based diagnostic technologies. In Artificial Intelligence, Machine Learning is one of the divisions which contains different types of statistical, probabilistic, and optimization techniques, allows machines to learn from past experience, and has the ability to detect different

DOI: 10.1201/9781003038450-10

types of patterns from the dataset. However, Machine Learning is not able to solve the problems related to complex datasets. For this purpose, Deep Learning is used for the recognition and analysis of medical images. For building the new Deep Learning tools with relative ease, scientists and researchers made the presence of open-source software possible (Abadi et al., 2016; Saba et al., 2019; Vedaldi et al., 2014). Today, most Deep Learning applications are better than humans at identifying the indicators for blood cancer and tumors present in MRI Scans. Deep Learning with Artificial Neural Networks consists of many hidden layers that allow for a higher level of abstraction and for improving the analysis of an image. Deep Neural Networks use various layers of neurons for the hierarchical representation of features. The accuracy of a diagnosis depends on the two processes:

Acquisition of Image: The creation of an object that is a digitally encoded representation of the visual characteristics.

Image Interpretation: It consists of the measurement of the object images, identification of the image, and proper use of the information in the problem.

This chapter focused on the clinical picture of handling different strategies for detecting various diseases using Deep Learning techniques. In medical services, Convolutional Neural Networks give better outcomes for training and testing information. Deep Learning techniques support the Convention Machine Learning algorithms because automatic feature extraction and selection processes are included. Generally, Deep Learning can be used in high-level applications, such as radiation therapy, CT-examined highlights extraction, MRI-SPECT check lessening remedy, and so on. Deep Learning applications in the field of CT scans and MRI are used for the recovery of a picture and division to predict infection in the beginning phases. CT scan is used to diagnose the disease by producing a number of images.

Capsule Endoscopy is used to examine the digestive tract of the human. Convolutional Neural Networks are used to train the Deep Learning-based image to differentiate the cancerous images. A study (Wang et al. 2020) used Deep Learning as a biomarker for pathological images.

Optical Coherence Tomography Images with Deep Convolutional Networks are used to improve the image quality of retinal images. Optical Coherence Tomography also provides a microscopic view of the retina. Deep Learning-based Neural Networks identify the areas of interest in the Optical Coherence Tomography Images.

10.1.1 CHALLENGES OF MEDICAL IMAGE PROCESSING

1. There are limited datasets of available high-quality medical images. These techniques do not depend much on medical image information, but they need efficient data. To create the data, medical experts are required and it is a time-consuming process. In the medical field, most of the data is unbalanced because data comes from the opinions of multiple medical experts. Deep Learning requires well-balanced information for training and testing.

2. Presently, most of the Deep Learning models are trained with simple 2-dimensional images (conventional x-ray images). However, CT scans and MRI scan images are usually 3-dimensional images. Adding extra features to

the Deep Learning models is very complicated because the machine has to be trained with 3D medical images.

3. The acquisition of non-standardized data is becoming the most challenging issue in the field of medical image processing. As a large number of various disease data becomes available, a huge number of datasets are needed for training the Deep Learning models.

4. As per the Health Insurance Portability and Accountability Act of 1996, medical experts cannot share personal records with anyone because the comparison of one dataset with others is a crucial and difficult task. The information related to all diseases should be kept in HIPAA, which means that there are privacy and legal issues to accessing this private data.

10.2 DEEP LEARNING

Deep Learning is a part of man-made reasoning and relies on Artificial Neural Networks (ANN). Neural Networks are a copy of the human mind with the number of neurons or hubs. Deep Learning comprises various applications in the field of well-being, including tumor identification (Liu et al., 2014), following of tumor advancement (Kather et al., 2019; Wang et al., 2020), bloodstream representation (Choi et al., 2017), diabetic retinopathy (Dutta et al., 2018; Gargeya & Leng, 2017), and so on. A few Deep Learning models are used to detect different infections, including Convolutional Neural Networks (CNN), Autoencoders, Transfer Learning, Deep Belief Networks (DBN), Recurrent Neural Networks (RNN), Generative Adversarial Networks (GAN), and Restricted Boltzmann Machine (RBM) learning calculations (Figure 10.1).

10.2.1 IMPORTANT DEEP LEARNING MODELS

10.2.1.1 Supervised Learning Models

In supervised learning, the machine is trained with the labeled class information that implies the past input-output pairs. Several types of supervised learning algorithms

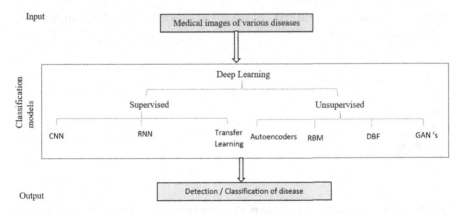

FIGURE 10.1 Deep learning algorithms used for object classification.

are used, such as Convolutional Neural Networks, Recurrent Neural Networks, and Transfer Learning.

10.2.1.1.1 Convolutional Neural Networks (CNNs)

For analyzing medical images, Convolutional Neural Networks have become the best methodology (Litjens et al., 2017). While performing the dimensionality reduction, CNNs are extremely useful algorithms for image recognition and visual learning tasks (Anwar et al., 2018) because of their unique structure in storing the image relations locally. It is observed that CNNs coming from Deep Learning are computationally quite effective (Maier et al., 2019). The best advantage of these networks is to store the spatial relationships of given input images while performing the filtering operation on an image. Generally, CNNs consist of a number of layers, which include the convolutional layer, ReLU, pooling, and fully connected layer. Milletari et al. (2016) presented a 3D segmentation using Convolutional Neural Networks for finding the prostate from the MRI volumes (Figure 10.2).

Convolutional Layer: This layer is the heart of CNN. Generally, convolution depends on the two functions: input values, which are the number of pixels; and the filter, which is also called a Kernel. Both functions can be shown with the array of values or matrices. The output can be achieved with the dot product of array values of two functions. In the next step, the filter is moved to the next position of the image, called a stride length. Then, the same computation is repeated until the total image is covered. Finally, it delivers the element or enactment map. This guide shows where the channel is firmly enacted for the given picture that implies solid highlights. These highlights are given as contributions to the ReLU. The 1D and 2D convolution tasks (Ker et al., 2017) appear underneath. This convolutional layer operation is represented by the * or . (dot) symbol:

$$S(t) = (I * K)(t)$$

where S(t) represents output or feature map, I(t) represents input or matrix, K(a) represents Filter or Kernel, and t represents an integer value.

The discretized convolution is
These operations are for the 1D convolutional operation. The 2D convolutional operation is as follows:

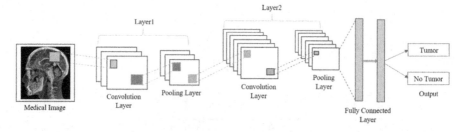

FIGURE 10.2 A general representation of convolutional neural network in healthcare applications.

When Input is I(m, n) and Filter is K(a, b), the convolution operation is

$$S(t) = \sum_a \sum_b I(a, b). \, K(m - a, n - b)$$

From commutative law, Filter is reversed.

$$S(t) = \sum_a \sum_b I(m - a, \, n - b). \, K(a, b)$$

Finally, neural networks support the kernel function.

$$S(t) = \sum_a \sum_b I(m + a, \, N + b). \, K(a, b)$$

Rectified Linear Unit (ReLU): It is the second layer of Convolutional Neural Networks. It performs threshold operations to the given input array values. If any value is less than 0 (means negative values), then it automatically sets to 0. This layer simplifies the calculation.

$$f(x) = \max(0, x)$$

where x is input to the neuron.

Pooling Layer: This layer is found between the convolution and ReLU layers. It is used to decrease calculations and the width and height of the given image but not the depth. The most common technique is max-pooling. This pooling takes the largest value in the filters and ignores the remaining values.

Fully Connected Layer: In this layer, all the input values are connected to every activation feature which is presented in the next layer. Fully connected means every neuron takes the input from every element of the previous layer. Depending upon the required feature abstraction, one or more fully connected layers are used.

10.2.1.1.2 Recurrent Neural Networks

Recurrent Neural Networks handle the single length or sequential input or output lengths. They use internal memory to store the sequence of inputs. Despite all traditional neural networks, RNNs are completely different because all the inputs and outputs are independent of each other. RNNs are mostly used in speech recognition and natural language processing, but in medical image analysis, RNNs are used for segmentation. To reduce the vanish gradient problems, RNNs are using the LSTM to hold the data for a long period. Generally, RNN models are used to construct the natural language descriptions for the images with their regions (Karpathy & Fei-Fei, 2015). A combination of CNN and RNN gives the best results for the longitudinal analysis of MRI Images of the brain in detecting Alzheimer's disease (Cui et al., 2019).

10.2.1.1.3 Transfer Learning

It is one of the popular learning tasks where the machine is trained with the data available in one particular task and again it is used for another task. Nowadays, this learning has become more popular because Deep Neural Networks can be trained to a little dataset as well. Transfer Learning is used for the classification of two different Computer-aided diagnosis-based disease classifications, thoracoabdominal lymph hub discovery and interstitial lung infection. Otitis Media is the infection located inside the ear. For the treatment of Otitis Media, Shie et al. (2015) presented the feature representation and scarcity of training data, using transfer learning (Figure 10.3) (Shie et al., 2015).

10.2.1.2   Unsupervised Learning

In solo learning, unlabeled data was used. These learning models are used to tackle intricate issues. Different solo learning calculations are Autoencoders, Restricted Boltzmann Machines, Deep Belief Networks, and Generative Adversarial Networks.

10.2.1.2.1 Autoencoders

It is one of the types of Artificial Neural Networks. Without using labeled data, autoencoders are able to learn features from the input data. Autoencoders are used to compress the input and again reconstruct the output from the compressed input. Thus, these models are used for dimensionality reduction.

10.2.1.2.2 Restricted Boltzmann Machines and Deep Belief Networks

Autoencoders have a number of advantages, which include feature detection, reduction of dimensionality, and output reconstruction. For most of the parts, in autoencoders, the output must be equal to the input data, which means that the number of neurons introduced in the information and yield is consistently the same.

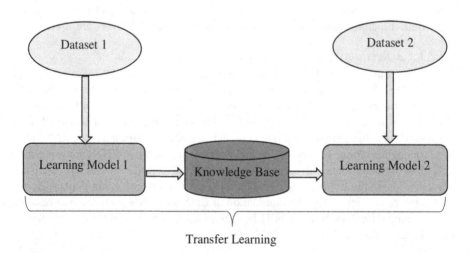

FIGURE 10.3   General architecture of transfer learning.

A stacked autoencoder contains several layers of autoencoders where the current output layer is connected to the next-state hidden layer. Some constraints are added to make the model learn for useful representations such as a denoising autoencoder, a sparse autoencoder, and variational autoencoders. For a content-based image retrieval challenge in medical image analysis, stacked autoencoders have given the best performance results with less error rate (Sharma et al., 2016). Convolutional Autoencoder Neural Network architecture was proposed for classifying the pulmonary nodules (Chen et al., 2017).

These networks are the two-layered Artificial Neural Networks. These machines can learn from a probability distribution over the set of input values. Restricted Boltzmann machines are the building blocks of Deep Belief Networks. The various steps of these networks are as follows: (1) variable or input, (2) hidden layer, and (3) a circle that represents a neuron. These two networks have the same representational power but the difference in the number of models represents the probabilistic distribution of data (Polania & Barner, 2017).

10.2.1.2.3 Generative Adversarial Networks

These algorithms are the Deep Learning-based generative model, which comprises two networks: a generator and a discriminator. In medical image analysis, Generative Adversarial Networks are used for data augmentation, data scarcity, and overfitting problems. Mahapatra et al. (2019) developed Progressive Generative Adversarial Networks that accept the low-resolution image as an input and give the high-resolution image as an output for the accurate diagnosis in visual image processing applications. For obtaining smooth medical image registration, Generative Adversarial Networks are used in retinal image registration results (Mahapatra et al., 2018). Various important Deep Learning architectures are shown in Figure 10.4.

10.3 MEDICAL IMAGE PROCESSING

It refers to the various techniques that are used to diagnose, detect, and treat medical conditions of the human body. Image processing refers to the processing of an image by applying different techniques, such as image enhancement, contrast enhancement, edge detection, and feature selection.

10.3.1 CARDIOVASCULAR DISEASES

As per the WHO in 2016, heart diseases are a major reason for death in the total population, including arrhythmia, coronary artery disease (CAD), and myocardial infarction. These are the causes of heart attacks or cardiac arrests. Electrocardiograms (ECGs) and angiograms are used for the primary diagnosis of heart diseases as prescreening and physical examination of the heart.

10.3.2 ARRHYTHMIA

Arrhythmia means an improper heartbeat so that the functionality of the heart decreases day by day. Cai et al. (2020) used the Convolutional Neural Networks for

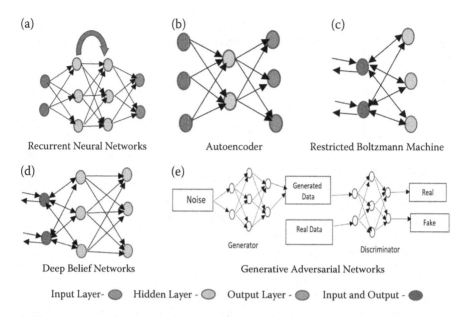

FIGURE 10.4   Various important deep learning models.

computer vision tasks, ECG detection, and 12-lead ECG used for the convolution of 12 channels. Also, this study proposed a multi-ECGNet for improving the exhibition of the location of ECG. This multi-ECGNet identifies all heart diseases at a time using the Deep Learning method for the classification of multi-labels.

Gul et al. (2017) focused on the 18 types of arrhythmia class classification. Preprocessing is applied to remove the noise from the ECG signal. For adding the features, Discrete Wavelet Transform is used for the morphological feature. RR-interval and Teager Energy Operator are used for the dynamic features of ECG. For the proper balance of features, Independent Component Analysis is used for reducing features. To classify the different types of arrhythmia in ECG signal, Feed Forward Neural Networks are used along with Back Propagation.

10.3.3 CORONARY ARTERY DISEASE

CAD happens due to the damage in the heart's major blood vessels, the blockage of arteries. Cholesterol deposits in the arteries cause a reduction in the blood supply to the heart. Myocardial infarction, cardiac arrest, and heart failure are the types of CAD.

Jafarian et al. (2020) implemented the two methods for detecting myocardial infarction, including the exemplary multi-lead ECG handling framework and the start to finish profound Neural Networks. Discrete Wavelet Transform and Principal Component Analysis are used to extract the features from the 12-lead ECGs. To diagnose the infarcted areas of myocardium, a two-class Neural Network classifier was designed. Shallow Neural Networks and End-to-End Deep Residual Learnings are used to detect and localize the myocardial infarction.

Acharya et al. (2019) developed a mechanized PC-helped framework to analyze ECG signals in congestive heart failure. Eleven-Layer Deep Convolutional Neural Network models are used for distinguishing congestive heart failures and it characterizes the ECG signals into the normal signs and congestive heart failure classes. For this reason, four distinct models of datasets are used to prepare and test Deep Convolutional Neural Networks.

Sujadevi et al. (2019) used a phonocardiogram to record the heart sounds. Deep Learning calculations are used to distinguish the strange hints of the heart. They utilized diverse Deep Learning designs, such as Recurrent Neural Networks, Long Short-Term Memory, Bidirectional-Recurrent Neural Networks, Bidirectional Long Short-Term Memory, and Convolutional Neural Networks, to classify the heart sounds that are ordinary or anomalous. Out of every one of these calculations, CNNs gave the best order results (Figure 10.5).

10.4 PARKINSON'S AND ALZHEIMER'S DISEASES

Parkinson's disease is the messiness of the cerebrum that causes shaking, firmness, strolling trouble, body equilibrium, and coordination. An electro-encephalogram (EEG) is used for the early detection of mind variations from the norm. Oh et al. (2018) used the EEG signal data of 20 Parkinson's patients and they proposed the 13-layer CNN model for the identification of Parkinson's disease from EEG signals.

Generally, SPECT scan is used to test the blood flow in other tissues and organs. Choi et al. (2017) developed the Deep Learning-based Fluro Propyl – carbomethoxy

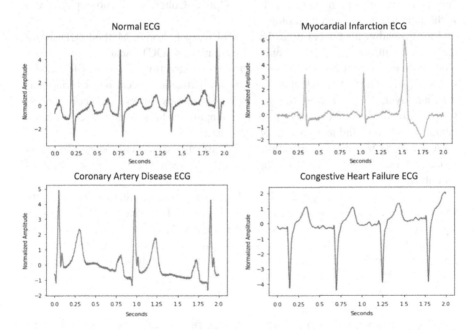

FIGURE 10.5   Normal ECG versus different types of heart disease ECGs (Lih et al., 2020).

iodophenyl nor propane – SPECT imaging test was used for the diagnosis of Parkinson's disease.

Alzheimer's disease is a disorder of the neurological system caused by abnormal proteins where amyloid grows in and around brain cells. Liu et al. (2014) designed a model with Deep Learning methods, consisting of stacked autoencoders and a SoftMax output layer used for the detection or diagnosis of Alzheimer's disease.

Suk and Shen (2013) proposed a Deep Learning-based representation of features with a stacked autoencoder. They performed several experiments on ADNI Dataset for the classification as Alzheimer's disease, Mild Cognitive Impairment (MCI), and MCI- Converter diagnosis.

10.4.1 Eye Diseases

The different eye disorders include glaucoma, age-related macular degeneration, diabetic retinopathy, and cataracts. Eye sicknesses prompt harm and visual impairment of the eye if they are not dealt with ahead of the schedule. Sometimes hemorrhages can occur when a small blood vessel breaks the bottom of the clear surface of the conjunctiva. In some situations, these hemorrhages may lead to the damage of an eye. Van Grinsven et al. (2016) proposed an approach for improving and speeding up the training of Convolutional Neural Networks by randomly selecting the improperly classified negative samples to detect the hemorrhages using the color fundus images.

Glaucoma is an eye disease that occurs due to high pressure on the eye that causes optic nerve damage, which is located at the back of the eyeball. Glaucoma can be the cause of permanent vision loss. Optical Coherence Tomography (OCT) is the imaging test in ophthalmology.

Muhammad et al. (2017) used the Hybrid Deep Learning Method (HDLM) to access the information from only one patient's OCT scan image. Later, Convolutional Neural Networks are used for the extraction of features from OCT scans. Chen et al. (2015) adopted response-normalization layers and overlapping-pooling layers for reducing the overfitting problems of eye glaucoma. Deep Convolutional Neural Networks are used for capturing the variant images for better characterization of hidden patterns of glaucoma.

Age-related Macular Degeneration usually affects the retina of the eye. When the retina is damaged, it is called macular damage. It mostly occurs in people aged 60 and above.

Peng et al. (2019) designed a Deep Learning model to characterize patients' shading fundus images using age-related eye illnesses. Treder et al. (2018) proposed a computerized distinguishing proof of Age-related Macular Degeneration in the Spectral Domain-OCT utilizing Deep Learning techniques. ImageNet was pre-trained with the SD-OCT images of 1,012 patients; out of those, 701 patients have Age-related Macular Degeneration and 311 are healthy.

Diabetic Retinopathy is a diabetes problem that harms the veins of the eye's retina with hypertension sugar levels.

Gargeya and Leng (2017) based on the color fundus images of diabetic patients classified the data as healthy and diabetic retinopathy. In total, 75,137 color fundus images are used for training and testing with an accuracy of 97%. The area under the receiver-operating characterstic curve is used to calculate the precision. Dutta et al. (2018) trained the color fundus images data with Back Propagation, Deep Neural Networks, and Convolutional Neural Networks, and tested with the FUNDUS dataset. Deep Neural Networks gave the best performance.

If melanoma pigments are developed around or inside of the eye, it is called an Ocular Melanoma. Eye or Ocular Melanoma is very difficult to detect because abnormal tissues are developed under the retina.

Ganguly et al. (2019) introduced a robotized Eye Melanoma order using Convolutional Neural Networks. One hundred and seventy pre-analyzed examples are gathered from clinical preliminaries and the achieved accuracy is 91.78%.

10.4.2 Breast Cancer

Breast cancer is a typical intrusive malignant growth in people formed in the compartments called breast cells. The survival chances depend on the staging and treatment of the cancer. Treatments are chemotherapy, radiation, hormone therapy, and surgery. In advanced stages, the abnormal breast cells spread to the other parts of the body.

Hoque et al. (2019) used ultrasound grayscale images for the detection of breast cancer. They identified that green and blue channels are needed for the segmentation instead of red, green, and blue. OTSU's threshold-based segmentation techniques were used for segmentation and the achieved accuracy is 93.11%. Khan et al. (2019) developed a Deep Transfer Learning framework for breast cancer detection and classification using breast cytology images. Picture highlights are separated from pre-prepared Convolutional Neural Network models.

Medical experts and doctors suggest mammography for diagnosing breast cancer in the early stages. It is like a low-dose x-ray. Li (2019) proposed DUALCORENET (DUAL-path COnditional REsidual NETworks) for mammography segmentation and classification. Celik et al. (2020) proposed multipurpose image analysis software for improving the diagnostic accuracy of the mammography images in breast cancer detection. Deep Transfer Learning with Artificial Neural Networks was used to identify pathological patterns in mammography.

Becker et al. (2017) worked on one subtype of breast cancer, Invasive Ductal Carcinoma (IDC). They used Deep Learning models for automatic detection of IDC and the BreakHis public dataset for the classification of malignant and benign classes of breast tumors.

10.4.3 Gastrointestinal Diseases

The gastrointestinal tract is from the mouth to the anus in between the salivary organs, pharynx, throat, stomach, pancreas, liver, gall bladder, small intestine system, and large intestine system are presented. Mainly, digestive diseases are disorders of the gastrointestinal tract. Immunotherapy is used for the treatment of digestive diseases, which builds the body's immunity to fight against cancer tissues. Wang et al. (2020) used Deep Learning methods for predicting gastrointestinal cancer, such as gastric cancer and colon cancer, from the pathology images of immunotherapy. They achieved the information status in the form of the Tumor Mutation Burden.

Kather et al. (2019) prepared a Convolutional Neural Network with profound leftover learning for the arrangement of tumors. Microsatellite Instability was used to determine the response of patients to immunotherapy. Weston et al. (2019) proposed the fully automated 2D segmentation algorithm for the abdominal CT Scan images to quantify the body decomposition analysis on several patients using a Deep Learning approach. For segmenting the abdomen of CT, a Convolutional Neural Network with U-Net architecture trained with the 2,430 2D CT images.

In the early stages, for diagnosing gastric problems, an endoscopy is used to examine the digestive tract of a patient. It is a non-surgical procedure. Shibata et al. (2020) proposed a method for the detection of early gastric cancer from endoscopic images of the gastrointestinal tract using the Mask Recurrent-Convolutional Neural Networks method. They collected 1,208 healthy samples and 533 cancer images, applied Recurrent-Convolutional Neural Networks and U-Net architecture for comparing the results, and found that Recurrent-Convolutional Neural Network is best.

Lee et al. (2019) used Deep Neural Networks for classifying the ulcer as a benign or ulcer from color endoscopy images. The data collected from the Gil Hospital comprises 200 normal, 367 cancerous, and 220 ulcer cases and the achieved accuracy is >90% (Figure 10.6).

10.4.4 SKIN CANCER

Skin Cancer is, for the most part, classified as non-melanoma and melanoma. Non-melanoma malignant growths, such as basal cell carcinoma and squamous cell carcinomas, are not harmful and infrequently are spread to different parts of the body. This cancer is generally reparable whenever recognized in the early stages. Melanoma is the most forceful sort of skin disease, brought about by melanocytes, which create skin tone in the human body. The irregular cells of the melanoma can be spread to different organs of the body within six weeks.

Al Nazi and Abir (2020) presented a technique for the division of skin injury and melanoma location using Deep Learning algorithms. A U-Net framework with spatial dropout gives the best results of classification. Johansen et al. (2020) used hyperspectral images for the classification of skin lesions. High-quality dataset images are captured from both hyperspectral and conventional RGB images. Dorj et al. (2018) addressed the classification of skin lesions as benign or malignant using the ECOC SVM, Deep Convolutional Neural Network algorithms. RGB images of skin lesions were collected from the Internet and public datasets (Figure 10.7).

Premaladha and Ravichandran (2016) proposed a Computer-Aided Detection System for the classification and early detection of skin cancer. A normalized segmentation algorithm, a Differentiation Limited Adaptive Histogram Equalization (CLAHE) method, and a middle channel are used for picture upgrading. Finally, Deep Learning-based Neural Networks and Hybrid Adaboost SVM calculations are

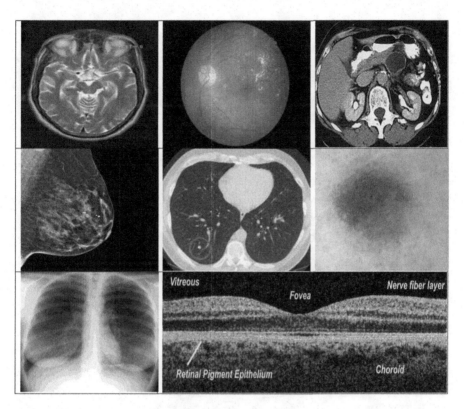

FIGURE 10.6 Left to Right: (i) CT image of brain lesions, (ii) diabetic retinopathy, (iii) MRI of gastro-intestinal disease, (iv) Mammographic image of breast, (v) CT of lung cancer, (vi) Dermoscopic image of skin, (vii) Chest X-ray of lung cancer, (viii) OCT of eye.

utilized for order. They physically gathered and tried 992 pictures from the clinical preliminaries and tested them.

Neural Networks cannot tackle the information loss and detection of the precise division of the boundary area. Zhang (2017) utilized a Neural Network structure for the analysis of skin disease to decrease screening errors. A batch normalization layer is added between the convolutional layer and activation layer to tackle the information loss and gradient disappearance.

Tan et al. (2019) proposed a Decision Support System for the discovery of skin malignant growth. Molecule Swarm Intelligence has been added for the element enhancement. Pathan et al. (2018) designed a precise methodology for the recognition of shade organization. Two-layer Feed Forward Neural Networks are utilized for the separation of run-of-the-mill and atypical organization designs. Seeja and Suresh (2019) improved the presentation of order using Deep Learning-based skin sore division. For removing the shape, surface, shading, and highlighting, various strategies are utilized and the accomplished grouping precision is 85.19%.

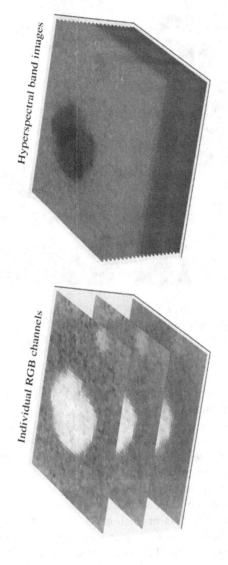

FIGURE 10.7 Conventional RGB image versus hyperspectral band image.

10.4.5 LIVER DISEASES

In the human body, the liver is an important organ that digests food and converts it into energy. The liver is located in the upper right part of the stomach and it is a football-sized organ. There are many liver diseases, including hepatitis, fatty liver disease, autoimmune conditions, genetic conditions, cancer, cirrhosis, and liver failure. The diagnosis methods of liver diseases are the liver function test, a complete blood count test, a CT Scan, an MRI for liver damage, and a liver biopsy for diagnosing cancer.

Fatty Liver Disease is brought about by the fat development in the liver. It occurs regularly in individuals who are corpulent and overweight. It is divided into two types: one is alcoholic fatty liver sickness and the other is non-alcoholic fatty liver disease (NAFLD). Generally, fat develops in the liver since liquor in NAFLD fat creates in individuals who do not drink a lot of liquor. Cao et al. (2020) collected information from 240 patients and split them into classes called typical, mild, tweak, and extreme NAFLD. They utilized a 2D ultrasound imaging dataset for the utilization of deep learning in quantitative examination histogram. The area under curve values give the best performance results for image processing methods. Another study (Reddy et al., 2018), for recognizing the fatty liver illness from the ultrasound images, built up one Computer-Aided Design model with deep learning, fine tuning, and transfer learning models for the classification of fatty liver sickness. Later on, another investigation (Byra et al., 2018), for calculating the measure of fat in the liver, proposed a Neural Network-based technique for NAFLD from ultra-sound images. They utilized the Inception-ResNet-V2 profound convolutional neural organization to remove the undeniable level highlights from the liver. This technique may be very valuable for sonographers.

Chronic liver disease is caused by hepatitis, other viruses, and the abuse of al-cohol. A chronic liver disease diagnosis can be improved later by the examination study, applied Convolutional Neural Networks for better results. For this purpose, Shear Wave Elastography (SWE) images were used for the automatic detection of chronic liver disease, and it avoids unnecessary measurements in unreliable areas (Gatos et al., 2019).

When abnormal cells are formed in hepatocytes, it is called heptachlor carci-noma, another term for liver cancer. The symptoms of this carcinoma are weight loss, stomach pain, vomiting, and change in skin color. The treatments of this cancer may vary at different stages and can include the removal of some parts, transplantation, chemotherapy, and radiation. A group of abnormal cells in the liver is called a liver lesion, tumor, or mass. Yasaka et al. (2018) investigated the diagnostic performance of the liver masses with the help of CNNs and Deep Learning methods. For the diagnosis, they used the enhanced CT image data and split it into three categories: non-contrast agent-enhanced, arterial, and delayed. Finally, the data identified as arterial and delayed give the best results for the classification of liver masses. Roth et al. (2015) applied the Convolutional Networks to the CT images for computing the image features from different or-gans in the dataset, such as the neck, liver, lungs, pelvis, and legs in the classi-fication phase. From this dataset, 80% of the data is for training and 20% is for testing; they improved the area under the curve values. Arjmand et al. (2019) used the Deep Learning approaches for finding the presence of various liver biopsies in the liver with a classification accuracy of 95%. Trivizakis et al. (2018) proposed a 3D CNN for the classification of tissue and they tested with 130 diffusion-weighted MRI images. Das et al. (2019) implemented a new approach, called Watershed Gaussian-based Deep Learning, for a viable arrangement of a diseased tumor in the CT scan and they accomplished a classification accuracy of 99.38%.

10.4.6 LUNG CANCER

Lung cancer is becoming the most dangerous cancer which develops in the lungs. The tests that are used for the diagnosis of lung cancer are CT scans and chest x-rays.

Shakeel et al. (2019) proposed a method for improving the quality of lung images by reducing misclassification. For the removal of noise from an image, the weighted mean histogram equalization approach is used. Deep Learning instantaneously trained the Neural Network used for the classification of an image as cancerous or non-cancerous.

The primary disease that affects the lungs is tuberculosis. For non-invasive diagnosis and screening, chest x-rays are useful. Nguyen et al. (2019) improved the performance of tuberculosis classification using Deep Learning methods. ImageNet was used and trained with a multi-label multi-class scenario because low-level features of ImageNet were not used for x-rays.

Gunasinghe et al. (2019) observed that Convolutional Neural Networks have extra highlights for anticipating cellular breakdown in the lungs contrasted with container organization. Lakshmanaprabu et al. (2019) proposed an Optimal Deep Neural Network with the feature reduction utilized for better characterization of lung CT pictures. Deep belief networks and restricted Boltzmann machine algorithms are used for the characterization of lung images. Huo et al. (2019) used a Neural Network model with a U-Net structure for segmenting the lung region of CT (Table 10.1).

10.5 CONCLUSION

Nowadays, Deep Learning in medical image processing is going on with more research and development. Computer-Aided Diagnosis in medical image analysis reduces the surgeries. Exact detection of objects or tumors can be done using very few sample datasets. Mostly, many researchers have worked with semi-supervised and unsupervised Deep Learning architectures only. In this chapter, various diseases are compared with the different types of Deep Learning techniques. Further, based on their outstanding performance, Deep Learning methods will overcome the humans in the sector of medical image processing for identifying and diagnosing an object. Extensive experiments are required on the state-of-the-art applications of Deep Learning for reducing the barriers in medical image processing.

TABLE 10.1

Comparison of Various Deep Learning Techniques in Medical Image Processing with Accuracies

Reference	Name of Disease	Type of Disease	Image Type	Deep Learning Algorithm Used	Data Set	Accuracy
Gul et al. (2017)	Heart Disease	Arrhythmia	ECG signal images	Independent Component Analysis, Feed Forward Neural Networks	MIT-BIH	99.4%
Martinez-Velazquez et al. (2019)		Cardiovascular Disease		Convolutional Neural Networks	PTB Diagnostic ECG Data	85.77%
Lih et al. (2020)		Cardiovascular Disease		Combined CNN with Long Short-Term Memory (LSTM)	PTB Diagnostic ECG Data	98.51%
Acharya et al. (2019)		Congestive Heart Failure		11-Layer Deep Convolutional Neural Networks	Physio Bank	98.97%
Jafarian et al. (2020)		Myocardial Infarction		Deep Residual Networks, Artificial Neural Networks	PTB Dataset	98.21%
Sujadevi et al. (2019)		Phonocardiogram		Recurrent Neural Networks, Long Short-Term Memory, Convolutional Neural Networks	PhysioNet-2016	CNN is best
Godkhindi and Gowda (2017)		Colon Cancer	CT images	Convolutional Neural Networks	TCIA Dataset	87.03 (Colon cancer) 88.56 (Polyp detection)
Ding et al. (2019)	Gastrointestinal cancer	Small Bowel Cancer	Capsule endoscopy images	Deep- Convolutional Neural Networks	113426569 Images from 6970 Patients	74.57%
Wang et al. (2020)		Gastric and Colon cancers Tumor Mutation Burden	Histology images	Convolutional Neural Networks based Transfer Learning	FFPE Dataset	77%
Weston et al. (2019)		Abdomen segmentation	CT images	Convolutional Neural Networks based on U-Net Architecture	2707 images from 1429 patients	–
Shibata et al. (2020)		Gastric Cancer	Endoscopy images	Mask Recurrent- Convolutional Neural Networks	1208 healthy and 533 cancer images from clinics	96%
Lee et al. (2019)		Gastric Cancer	Endoscopy images	Inception, ResNet, VGGNet	Clinical data from GIL Hospital	ResNet is best
Shakeel et al. (2019)	Lung Diseases	Lung Cancer	CT images	Deep Learning Instantaneously Neural Networks	CIA dataset	98.42%
Huo et al. (2019)		Lung Cancer	CT images	Convolutional Neural Networks, Artificial Neural Networks	770 chest x-ray images	–
Nguyen et al. (2019)		Lung Cancer	CT images	Optimal Deep Neural Networks, Linear Discriminant Analysis	50 low dosage lung CT images	96.2%
Lakshmanaprabu et al. (2019)		Tuberculosis	Chest x-ray images	Transfer Learning using ImageNet	X-rays from NH-14	–
Al Nazi and Abir (2020)	Skin Diseases	Skin Cancer	Dermoscopy images	Convolutional Neural Networks based on U-Net Architecture with SVM	PH2 and ISIC-2018	92%

(Continued)

TABLE 10.1 (Continued)

Comparison of Various Deep Learning Techniques in Medical Image Processing with Accuracies

Reference	Name of Disease	Type of Disease	Image Type	Deep Learning Algorithm Used	Data Set	Accuracy
Dorj et al. (2018)		Skin Cancer	RGB images	Deep Convolutional Neural Networks	ALL-IDB2	–
Premaladha and Ravichandran (2016)		Melanoma	Dermoscopy images	ECOC SVM and Deep Convolutional Neural Networks	3753 skin lesion images	90.74%
Tan et al. (2019)		Melanoma	Dermoscopy images	Deep Learning-based Neural Networks, Hybrid AdaBoost-SVM	992 skin lesion images	93%
Pathan et al. (2018)		Melanoma	Dermoscopy images	Probabilistic SVM, Gray Level Co-occurrence Matrix	PH2 Dataset of 200 images	96.7%
Seeja and Suresh (2019)		Melanoma	Dermoscopy images	Convolutional Neural Network with U-NET, SVM, Random Forest, K-Nearest neighbors, Naïve Bayes	ISBI-2016 Dataset	SVM is best
Muhammad et al. (2017)	Eye Diseases	Glaucoma	OCT scan images	Hybrid Deep Learning Method, Convolutional Neural Networks	102 eyes from 102 patients	87.3%
Chen et al. (2015)		Glaucoma	Digital fundus images	Convolutional Neural Networks	ORIGA SCES Data	83.1%88.7%
Treder et al. (2018)		Age-related Macular Degeneration	SD-OCT images	Deep- Convolutional Neural Networks	1012 Spectral Domain OCT Images	99.7%
Gargeya and Leng (2017)		Diabetic Retinopathy	Color fundus images	Deep Residual Learning with Convolutional Neural Networks	75137 CFI from EyePACS Dataset	97%
Dutta et al. (2018)		Diabetic Retinopathy	Color fundus images	Deep Neural Networks, Convolutional Neural Networks	2000 images from Kaggle	72.5%
Ganguly et al. (2019)		Ocular Melanoma	Dermoscopy, color fundus images	Convolutional Neural Networks, Artificial Neural Networks	170 pre-diagnosed samples	91.76%
Oh et al. (2018)	Brain Diseases	Parkinson's Disease	EEG signal images	13-Layer Convolutional Neural Networks	40 patients EEG signals	88.25%
Choi et al. (2017)		Parkinson's Disease	SPECT scan images	Deep Neural Networks	PPMI Database	92%
Liu et al. (2014)		Alzheimer's Disease	MRI/PET scan images	Deep Learning with autoencoders and SoftMax Regression layer	ADNI Dataset	87.76%
Suk and Shen (2013)		Alzheimer's Disease (AD)/ Mild Cognitive Impairment (MCI)	MRI/PET scan images	Deep Learning with autoencoders	ADNI Dataset	AD: 95.9% MCI: 85%
Reddy et al. (2018)	Liver Diseases	Fatty Liver Disease (FLD)	Ultrasound images	Convolutional Neural Networks and Transfer Learning	Collected from sonographers	90.6%

REFERENCES

Abadi, M., Barham, P., Chen, J., Chen, Z., Davis, A., Dean, J., & Kudlur, M. (2016). Tensorflow: A system for large-scale machine learning. In *12th {USENIX} Symposium on Operating Systems Design and Implementation ({OSDI} 16)* (pp. 265–283).

Acharya, U. R., Fujita, H., Oh, S. L., Hagiwara, Y., Tan, J. H., Adam, M., & San Tan, R. (2019). Deep convolutional neural network for the automated diagnosis of congestive heart failure using ECG signals. *Applied Intelligence*, *49*(1), 16–27.

Al Nazi, Z., & Abir, T. A. (2020). Automatic skin lesion segmentation and melanoma detection: Transfer learning approach with U-NET and DCNN-SVM. In *Proceedings of International Joint Conference on Computational Intelligence* (pp. 371–381), Singapore.

Anwar, S. M., Majid, M., Qayyum, A., Awais, M., Alnowami, M., & Khan, M. K. (2018). Medical image analysis using convolutional neural networks: A review. *Journal of Medical Systems*, *42*(11), 226.

Arjmand, A., Angelis, C. T., Tzallas, A. T., Tsipouras, M. G., Glavas, E., Forlano, R., & Giannakeas, N. (2019, July). Deep learning in liver biopsies using convolutional neural networks. In *2019 42nd International Conference on Telecommunications and Signal Processing (TSP)* (pp. 496–499).

Becker, A. S., Marcon, M., Ghafoor, S., Wurnig, M. C., Frauenfelder, T., & Boss, A. (2017). Deep Learning in mammography: Diagnostic accuracy of a multipurpose image analysis software in the detection of breast cancer. *Investigative Radiology*, *52*(7), 434–440.

Byra, M., Styczynski, G., Szmigielski, C., Kalinowski, P., Michałowski, L., Paluszkiewicz, R., & Nowicki, A. (2018). Transfer learning with deep convolutional neural network for liver steatosis assessment in ultrasound images. *International Journal of Computer Assisted Radiology and Surgery*, *13*(12), 1895–1903.

Cai, J., Sun, W., Guan, J., & You, I. (2020). Multi-ECGNet for ECG arrhythmia multi-label classification. *IEEE Access*.

Cao, W., An, X., Cong, L., Lyu, C., Zhou, Q., & Guo, R. (2020). Application of deep learning in quantitative analysis of 2-dimensional ultrasound imaging of nonalcoholic fatty liver disease. *Journal of Ultrasound in Medicine*, *39*(1), 51–59.

Celik, Y., Talo, M., Yildirim, O., Karabatak, M., & Acharya, U. R. (2020). Automated invasive ductal carcinoma detection based using deep transfer learning with whole-slide images. *Pattern Recognition Letters*, *133*, 232–239.

Chen, M., Shi, X., Zhang, Y., Wu, D., & Guizani, M. (2017). Deep features learning for medical image analysis with convolutional autoencoder neural network. *IEEE Transactions on Big Data*. doi: 10.1109/TBDATA.2017.2717439.

Chen, X., Xu, Y., Wong, D. W. K., Wong, T. Y., & Liu, J. (2015, August). Glaucoma detection based on deep convolutional neural network. In *2015 37th Annual International Conference of the IEEE Engineering in Medicine and Biology Society (EMBC)* (pp. 715–718).

Choi, H., Ha, S., Im, H. J., Paek, S. H., & Lee, D. S. (2017). Refining diagnosis of Parkinson's disease with deep learning-based interpretation of dopamine transporter imaging. *NeuroImage: Clinical*, *16*, 586–594.

Cui, R., Liu, M., & Alzheimer's Disease Neuroimaging Initiative. (2019). RNN-based longitudinal analysis for diagnosis of Alzheimer's disease. *Computerized Medical Imaging and Graphics*, *73*, 1–10.

Das, A., Acharya, U. R., Panda, S. S., & Sabut, S. (2019). Deep learning based liver cancer detection using watershed transform and Gaussian mixture model techniques. *Cognitive Systems Research*, *54*, 165–175.

Ding, Z., Shi, H., Zhang, H., Meng, L., Fan, M., Han, C., & Liu, J. (2019). Gastroenterologist-level identification of small-bowel diseases and normal variants by capsule endoscopy using a deep-learning model. *Gastroenterology*, *157*(4), 1044–1054.

Dorj, U. O., Lee, K. K., Choi, J. Y., & Lee, M. (2018). The skin cancer classification using deep convolutional neural network. *Multimedia Tools and Applications*, *77*(8), 9909–9924.

Dutta, S., Manideep, B. C., Basha, S. M., Caytiles, R. D., & Iyengar, N. C. S. N. (2018). Classification of diabetic retinopathy images by using deep learning models. *International Journal of Grid and Distributed Computing*, *11*(1), 89–106.

Ganguly, B., Biswas, S., Ghosh, S., Maiti, S., & Bodhak, S. (2019, January). A deep learning framework for eye melanoma detection employing convolutional neural network. In *2019 International Conference on Computer, Electrical & Communication Engineering (ICCECE)* (pp. 1–4).

Gargeya, R., & Leng, T. (2017). Automated identification of diabetic retinopathy using deep learning. *Ophthalmology*, *124*(7), 962–969.

Gatos, I., Tsantis, S., Spiliopoulos, S., Karnabatidis, D., Theotokas, I., Zoumpoulis, P., & Kagadis, G. C. (2019). Temporal stability assessment in shear wave elasticity images validated by deep learning neural network for chronic liver disease fibrosis stage assessment. *Medical Physics*, *46*(5), 2298–2309.

Godkhindi, A. M., & Gowda, R. M. (2017, August). Automated detection of polyps in CT colonography images using deep learning algorithms in colon cancer diagnosis. In *2017 International Conference on Energy, Communication, Data Analytics and Soft Computing (ICECDS)* (pp. 1722–1728).

Gul, M., Anwar, S. M., & Majid, M. (2017, October). Electrocardiogram signal classification to detect arrhythmia with improved features. In *2017 IEEE International Conference on Imaging Systems and Techniques (IST)* (pp. 1–6), China.

Gunasinghe, A. D., Aponso, A. C., & Thirimanna, H. (2019, March). Early prediction of lung diseases. In *2019 IEEE 5th International Conference for Convergence in Technology (I2CT)* (pp. 1–4), India.

Hoque, M. U., Sazzad, T. S., Farabi, A. A., Hosen, I., & Somi, M. A. (2019, May). An automated approach to detect breast cancer tissue using ultrasound images. In *2019 1st International Conference on Advances in Science, Engineering and Robotics Technology (ICASERT)* (pp. 1–4).

Huo, D., Kiehn, M., & Scherzinger, A. (2019). Investigation of low-dose CT lung cancer screening scan "over-range" issue using machine learning methods. *Journal of Digital Imaging*, *32*(6), 931–938.

Jafarian, K., Vahdat, V., Salehi, S., & Mobin, M. (2020). Automating detection and localization of myocardial infarction using shallow and end-to-end deep neural networks. *Applied Soft Computing*, 106383.

Johansen, T. H., Møllersen, K., Ortega, S., Fabelo, H., Garcia, A., Callico, G. M., & Godtliebsen, F. (2020). Recent advances in hyperspectral imaging for melanoma detection. *Wiley Interdisciplinary Reviews: Computational Statistics*, *12*(1), e1465.

Karpathy, A., & Fei-Fei, L. (2015). Deep visual-semantic alignments for generating image descriptions. In *Proceedings of the IEEE Conference on Computer Vision and Pattern Recognition* (pp. 3128–3137).

Kather, J. N., Pearson, A. T., Halama, N., Jäger, D., Krause, J., Loosen, S. H., & Grabsch, H. I. (2019). Deep Learning can predict microsatellite instability directly from histology in gastrointestinal cancer. *Nature Medicine*, *25*(7), 1054–1056.

Ker, J., Wang, L., Rao, J., & Lim, T. (2017). Deep learning applications in medical image analysis. *IEEE Access*, *6*, 9375–9389.

Khan, S., Islam, N., Jan, Z., Din, I. U., & Rodrigues, J. J. C. (2019). A novel deep learning based framework for the detection and classification of breast cancer using transfer learning. *Pattern Recognition Letters*, *125*, 1–6.

Lakshmanaprabu, S. K., Mohanty, S. N., Shankar, K., Arunkumar, N., & Ramirez, G. (2019). Optimal deep learning model for classification of lung cancer on CT images. *Future Generation Computer Systems*, *92*, 374–382.

Lee, C. S., Baughman, D. M., & Lee, A. Y. (2017). Deep learning is effective for classifying normal versus age-related macular degeneration OCT images. *Ophthalmology Retina*, *1*(4), 322–327.

Lee, J. H., Kim, Y. J., Kim, Y. W., Park, S., Choi, Y. I., Kim, Y. J., & Chung, J. W. (2019). Spotting malignancies from gastric endoscopic images using deep learning. *Surgical Endoscopy*, *33*(11), 3790–3797.

Li, H., Chen, D., Nailon, W. H., Davies, M. E., & Laurenson, D. (2019, May). A deep dual-path network for improved mammogram image processing. In *ICASSP 2019-2019 IEEE International Conference on Acoustics, Speech and Signal Processing (ICASSP)* (pp. 1224–1228).

Lih, O. S., Jahmunah, V., San, T. R., Ciaccio, E. J., Yamakawa, T., Tanabe, M., & Acharya, U. R. (2020). Comprehensive electrocardiographic diagnosis based on deep learning. *Artificial Intelligence in Medicine*, *103*, 101789.

Litjens, G., Kooi, T., Bejnordi, B. E., Setio, A. A. A., Ciompi, F., Ghafoorian, M., & Sánchez, C. I. (2017). A survey on deep learning in medical image analysis. *Medical Image Analysis*, *42*, 60–88.

Liu, S., Liu, S., Cai, W., Pujol, S., Kikinis, R., & Feng, D. (2014, April). Early diagnosis of Alzheimer's disease with deep learning. In *2014 IEEE 11th International Symposium on Biomedical Imaging (ISBI)* (pp. 1015–1018).

Mahapatra, D., Antony, B., Sedai, S., & Garnavi, R. (2018, April). Deformable medical image registration using generative adversarial networks. In *2018 IEEE 15th International Symposium on Biomedical Imaging (ISBI 2018)* (pp. 1449–1453).

Mahapatra, D., Bozorgtabar, B., & Garnavi, R. (2019). Image super-resolution using progressive generative adversarial networks for medical image analysis. *Computerized Medical Imaging and Graphics*, *71*, 30–39.

Maier, A., Syben, C., Lasser, T., & Riess, C. (2019). A gentle introduction to deep learning in medical image processing. *Zeitschrift für Medizinische Physik*, *29*(2), 86–101.

Martinez-Velazquez, R., Gamez, R., & El Saddik, A. (2019, June). Cardio Twin: A Digital Twin of the human heart running on the edge. In *2019 IEEE International Symposium on Medical Measurements and Applications (MeMeA)* (pp. 1–6).

Milletari, F., Navab, N., & Ahmadi, S. A. (2016, October). V-Net: Fully convolutional neural networks for volumetric medical image segmentation. In *2016 Fourth International Conference on 3D Vision (3DV)* (pp. 565–571).

Muhammad, H., Fuchs, T. J., De Cuir, N., De Moraes, C. G., Blumberg, D. M., Liebmann, J. M., ... Hood, D. C. (2017). Hybrid deep learning on single wide-field optical coherence tomography scans accurately classifies glaucoma suspects. *Journal of Glaucoma*, *26*(12), 1086.

Nguyen, Q. H., Nguyen, B. P., Dao, S. D., Unnikrishnan, B., Dhingra, R., Ravichandran, S. R., ... Chua, M. C. (2019, April). Deep learning models for tuberculosis detection from chest X-ray images. In *2019 26th International Conference on Telecommunications (ICT)* (pp. 381–385).

Oh, S. L., Hagiwara, Y., Raghavendra, U., Yuvaraj, R., Arunkumar, N., Murugappan, M., & Acharya, U. R. (2018). A deep learning approach for Parkinson's disease diagnosis from EEG signals. *Neural Computing and Applications*, 1–7.

Pathan, S., Prabhu, K. G., & Siddalingaswamy, P. C. (2018). A methodological approach to classify typical and atypical pigment network patterns for melanoma diagnosis. *Biomedical Signal Processing and Control, 44*, 25–37.

Peng, Y., Dharssi, S., Chen, Q., Keenan, T. D., Agrón, E., Wong, W. T., & Lu, Z. (2019). DeepSeeNet: A Deep Learning model for automated classification of patient-based age-related macular degeneration severity from color fundus photographs. *Ophthalmology, 126*(4), 565–575.

Polania, L. F., & Barner, K. E. (2017). Exploiting restricted Boltzmann machines and deep belief networks in compressed sensing. *IEEE Transactions on Signal Processing, 65*(17), 4538–4550.

Premaladha, J., & Ravichandran, K. S. (2016). Novel approaches for diagnosing melanoma skin lesions through supervised and deep learning algorithms. *Journal of Medical Systems, 40*(4), 96.

Reddy, D. S., Bharath, R., & Rajalakshmi, P. (2018, September). A novel computer-aided diagnosis framework using deep learning for classification of fatty liver disease in ultrasound imaging. In *2018 IEEE 20th International Conference on e-Health Networking, Applications and Services (Healthcom)* (pp. 1–5), IEEE.

Roth, H. R., Lee, C. T., Shin, H. C., Seff, A., Kim, L., Yao, J., & Summers, R. M. (2015, April). Anatomy-specific classification of medical images using deep convolutional nets. In *2015 IEEE 12th International Symposium on Biomedical Imaging (ISBI)* (pp. 101–104). IEEE.

Saba, L., Biswas, M., Kuppili, V., Godia, E. C., Suri, H. S., Edla, D. R., & Protogerou, A. (2019). The present and future of deep learning in radiology. *European Journal of Radiology, 114*, 14–24.

Seeja, R. D., & Suresh, A. (2019). Deep Learning based skin lesion segmentation and classification of melanoma using support vector machine (SVM). *Asian Pacific Journal of Cancer Prevention, 20*(5), 1555.

Shakeel, P. M., Burhanuddin, M. A., & Desa, M. I. (2019). Lung cancer detection from CT image using improved profuse clustering and deep learning instantaneously trained neural networks. *Measurement, 145*, 702–712.

Sharma, S., Umar, I., Ospina, L., Wong, D., & Tizhoosh, H. R. (2016, December). Stacked autoencoders for medical image search. In *International Symposium on Visual Computing* (pp. 45–54). Cham: Springer.

Shibata, T., Teramoto, A., Yamada, H., Ohmiya, N., Saito, K., & Fujita, H. (2020). Automated detection and segmentation of early gastric cancer from endoscopic images using mask R-CNN. *Applied Sciences, 10*(11), 3842.

Shie, C. K., Chuang, C. H., Chou, C. N., Wu, M. H., & Chang, E. Y. (2015, August). Transfer representation learning for medical image analysis. In *2015 37th Annual International Conference of the IEEE Engineering in Medicine and Biology Society (EMBC)* (pp. 711–714), IEEE.

Shin, H. C., Roth, H. R., Gao, M., Lu, L., Xu, Z., Nogues, I., & Summers, R. M. (2016). Deep convolutional neural networks for computer-aided detection: CNN architectures, dataset characteristics and transfer learning. *IEEE Transactions on Medical Imaging, 35*(5), 1285–1298.

Sujadevi, V. G., Soman, K. P., Vinayakumar, R., & Sankar, A. P. (2019). Anomaly detection in phonocardiogram employing deep learning. In *Computational Intelligence in Data Mining* (pp. 525–534). Singapore: Springer.

Suk, H. I., & Shen, D. (2013, September). Deep learning-based feature representation for AD/MCI classification. In *International Conference on Medical Image Computing and Computer-Assisted Intervention* (pp. 583–590), Berlin, Heidelberg.

Tan, T. Y., Zhang, L., & Lim, C. P. (2019). Intelligent skin cancer diagnosis using improved particle swarm optimization and deep learning models. *Applied Soft Computing, 84,* 105725.

Ting, D. S. W., Pasquale, L. R., Peng, L., Campbell, J. P., Lee, A. Y., Raman, R., ... Wong, T. Y. (2019). Artificial intelligence and deep learning in ophthalmology. *British Journal of Ophthalmology, 103*(2), 167–175.

Ting, D. S., Peng, L., Varadarajan, A. V., Keane, P. A., Burlina, P. M., Chiang, M. F., & Abramoff, M. (2019). Deep learning in ophthalmology: The technical and clinical considerations. *Progress in Retinal and Eye Research, 72,* 100759.

Treder, M., Lauermann, J. L., & Eter, N. (2018). Automated detection of exudative age-related macular degeneration in spectral domain optical coherence tomography using Deep Learning. *Graefe's Archive for Clinical and Experimental Ophthalmology, 256*(2), 259–265.

Trivizakis, E., Manikis, G. C., Nikiforaki, K., Drevelegas, K., Constantinides, M., Drevelegas, A., & Marias, K. (2018). Extending 2-D convolutional neural networks to 3-D for advancing deep learning cancer classification with application to MRI liver tumor differentiation. *IEEE Journal of Biomedical and Health Informatics, 23*(3), 923–930.

Van Grinsven, M. J., van Ginneken, B., Hoyng, C. B., Theelen, T., & Sánchez, C. I. (2016). Fast convolutional neural network training using selective data sampling: Application to hemorrhage detection in color fundus images. *IEEE Transactions on Medical Imaging, 35*(5), 1273–1284.

Vedaldi, A., Jia, Y., Shelhamer, E., Donahue, J., Karayev, S., Long, J., & Darrell, T. (2014). *Convolutional Architecture for Fast Feature Embedding.* Cornell University, *arXiv:1408.5093 v12014.*

Wang, L., Jiao, Y., Qiao, Y., Zeng, N., & Yu, R. (2020). A novel approach combined transfer learning and deep learning to predict TMB from histology image. *Pattern Recognition Letters, 135,* 244–248

Weston, A. D., Korfiatis, P., Kline, T. L., Philbrick, K. A., Kostandy, P., Sakinis, T., & Erickson, B. J. (2019). Automated abdominal segmentation of CT scans for body composition analysis using deep learning. *Radiology, 290*(3), 669–679.

Yasaka, K., Akai, H., Abe, O., & Kiryu, S. (2018). Deep learning with convolutional neural network for differentiation of liver masses at dynamic contrast-enhanced CT: A preliminary study. *Radiology, 286*(3), 887–896.

Yi, X., Walia, E., & Babyn, P. (2019). Generative adversarial network in medical imaging: A review. *Medical Image Analysis, 58,* 101552.

Zhang, X. (2017). Melanoma segmentation based on deep learning. *Computer Assisted Surgery, 22*(Suppl1), 267–277.

Index

Printed in the United States
by Baker & Taylor Publisher Services

Printed in the United States
by Baker & Taylor Publisher Services